Springer Geography

The Springer Geography series seeks to publish a broad portfolio of scientific books, aiming at researchers, students, and everyone interested in geographical research. The series includes peer-reviewed monographs, edited volumes, textbooks, and conference proceedings. It covers the entire research area of geography including, but not limited to, Economic Geography, Physical Geography, Quantitative Geography, and Regional/Urban Planning.

More information about this series at http://www.springer.com/series/10180

Jan Selmer Methi · Andrei Sergeev
Małgorzata Bieńkowska · Basia Nikiforova
Editors

Borderology: Cross-disciplinary Insights from the Border Zone

Along the Green Belt

Springer

Editors
Jan Selmer Methi
Faculty of Education and Arts
Nord University
Bodø, Nordland Fylke, Norway

Andrei Sergeev
Murmansk Arctic State University
Murmansk, The Murmansk Area, Russia

Małgorzata Bieńkowska
University of Białystok
Białystok, Poland

Basia Nikiforova
Lithuanian Culture Research Institute
Vilnius, Lithuania

ISSN 2194-315X ISSN 2194-3168 (electronic)
Springer Geography
ISBN 978-3-030-07597-2 ISBN 978-3-319-99392-8 (eBook)
https://doi.org/10.1007/978-3-319-99392-8

This Springer imprint is published by the registered company Springer Nature Switzerland AG
The registered company address is: Gewerbestrasse 11, 6330 Cham, Switzerland

Preface

This book is a collection of articles related to the ongoing project called Borderology. The project started in 2009 initiated by The Norwegian Barents Secretariat and was dedicated to build up education and research in order to develop new knowledge based on people-to-people-diplomacy and lived experience in the border zone between Norway and Russia. Nord university and Murmansk Arctic State University (MASU) became partners in this project and decided to develop a joint master program and arrange Bakhtin-Kant seminars annually on both sides of the border. The master program, Joint Master in Borderology, enrolled its first class in 2013. The seminars have resulted in six reports and one anthology, *Philosophy in the border zone* (2015). This book also represents the expanding of borderology thinking following the line of the Green Belt showing the collaborative project of an international group of authors which comprises, primarily, researchers from Nord University (Bodø, Norway) and Murmansk Arctic State University (Murmansk, Russia), as well as representatives of Saint Petersburg State University (St. Petersburg, Russia), the Lithuanian Culture Research Institute (Vilnius, Lithuania), Klaipėda University (Klaipėda, Lithuania), and the University of Białystok (Białystok, Poland). This is mirrored in the editorial board:

Andrei Sergeev is Doctor of Philosophy, Professor, full member of Russian Academy of Natural Sciences, rector of Murmansk Arctic State University. In 1984, he graduated from "History" training program of Petrozavodsk State University named after O. V. Kuusinen. In 1991–1993, he was enrolled in Ph.D. philosophy program of Leningrad State University. In 1997, he graduated from Higher Doctorate training in philosophy of Saint-Petersburg State University, and successfully passed a postdoctoral thesis defense on topic "Attitude to culture in the Russian metaphysical philosophy of XIX–XX century". In 1998, he was issued an academic degree Doctor of Philosophy. In 2001, he was issued a title of Professor. Spheres of academic interests: metaphysics, philosophy of culture, social philosophy.

Jan Selmer Methi is Associate professor in philosophy at Center for Practical knowledge, Nord university in Bodø, Norway. He completed his Ph.D. in philosophy at Århus University in Denmark in 2007. The topic of his Ph.D. was teacher's professional identity and self-awareness. His research interests include philosophy of science, sociocultural and activity theory, borderology and practical knowledge.

Basia Nikforova is Senior Research Fellow, Department of Contemporary Philosophy in Lithuanian Culture Research Institute. She completed her Ph.D. in philosophy in Moscow State University in 1977. Her interests of research include sociology of religion, cultural, and border studies, new materialism.

Małgorzata Bieńkowska is professor in sociology at Institute of Sociology and Cognitive Science, University of Bialystok, Poland and visiting professor University of Economy, Bydgoszcz, Poland. She is head of the Department of Sociology of Multiculturalism in Bialystok. She completed her Ph.D. in sociology at University of Nicolaus Copernicus in Torun, Poland. Her interests of research include theory of sociology, multiculturalism, anthropology of borderland, and gender studies.

Bodø, Norway
Murmansk, Russia
Białystok, Poland
Vilnius, Lithuania

Jan Selmer Methi
Andrei Sergeev
Małgorzata Bieńkowska
Basia Nikiforova

Contents

Part I Border Practices in the Green belt

On Uses of 'Wild Nature': Empowered Versus Disempowered Agency in Kola Reindeer Herding Territories (Murmansk Region, NW Russia) 3
Yulian Konstantinov

Borderlands of Lithuania and Kaliningrad Region of Russia: Preconditions for Comparative Geographic Approach and Spatial Interaction 15
Eduardas Spiriajevas

Movement to Defend the Białowieża—The Problem of the Białowieża Forest Protection as an Example of a Values Conflict 31
Małgorzata Bieńkowska, Łukasz Faszcza and Łukasz Wołyniec

Potential, Problems, and Challenges of Joint International Master Programmes: Case-Study of the Joint Norwegian-Russian Master Degree Programme in Borderology 41
Inna Ryzhkova and Jan Selmer Methi

Part II Challenges of Mental Borders: Challenge for Europe

The Power of Doubt 57
Boris Sokolov

Nature and Man: Crime and Punishment 69
Nadezhda Golik

Nature as Stoic 81
Viggo Rossvaer

What Role Plays Intuition in Mathematics and Science? On the Borders Between Several Conceptions of What It Means to Intuit 93
Johan Arnt Myrstad

Deconstruction of European Environmental Identity in the Mass Immigration Context Summary 121
Basia Nikiforova

Part III Knowledge on the Frontiers; Conceptualization of Practice (Activity)

Man and Nature: Approaches to the Delimitation of Concepts 135
Viktor R. Tsylev

Borderology and Practical Knowledge—Humanities Response to Epigenetics? 149
Jan Selmer Methi

Human Existence Between the Forest (Nature) and Home (Technology) 163
Vasilii M. Voronov

Beyond the Green Shift—Ecological Economics 173
Ove Jakobsen and Vivi M. L. Storsletten

Myth on the Border Between Man and Nature 185
Irina Gorina

Part IV Borders in Collision: Search of New Paradigms

Cultural Images of Nature as Paradigms of Human Practices 197
Aleksandr Sautkin

Imagination as a Breaker of the Border 207
Andrey Vinogradov

Part V New Unity in Science and Humanities: Renewal of Ethics

Translation in a Hermeneutical Context: Transforming Culture and Human Nature Through Ethics 217
Andrei Kopylov

The Need for Being Disinterested as a Key Characteristic of Human Nature 231
Andrei Sergeev

Index 247

Contributors

Małgorzata Bieńkowska Institute of Sociology and Cognitive Science, University of Białystok, Białystok, Poland

Łukasz Faszcza Institute of Sociology and Cognitive Science, University of Białystok, Białystok, Poland

Nadezhda Golik St. Petersburg State University, Saint Petersburg, Russia

Irina Gorina St. Petersburg State University, Saint Petersburg, Russia

Ove Jakobsen Nord University, Bodø, Norway

Yulian Konstantinov Bulgarian Society for Regional Cultural Studies, Sofia, Bulgaria

Andrei Kopylov Murmansk Arctic State University, Murmansk, Russia

Jan Selmer Methi Nord University, Bodø, Norway

Johan Arnt Myrstad Nord University, Bodø, Norway

Basia Nikiforova Lithuanian Culture Research Institute, Vilnius, Lithuania

Viggo Rossvaer Nord University, Bodø, Norway

Inna Ryzhkova Murmansk Arctic State University, Murmansk, Russia

Aleksandr Sautkin Murmansk Arctic State University, Murmansk, Russia

Andrei Sergeev Murmansk Arctic State University, Murmansk, Russia

Boris Sokolov St. Petersburg State University, Sankt-Peterburg, Russia

Eduardas Spiriajevas Klaipeda University, Klaipeda, Lithuania

Vivi M. L. Storsletten Nord University, Bodø, Norway

Viktor R. Tsylev Murmansk Arctic State University, Murmansk, Russia

Andrey Vinogradov Murmansk Arctic State University, Murmansk, Russia

Vasilii M. Voronov Murmansk Arctic State University, Murmansk, Russia

Łukasz Wołyniec Institute of Sociology and Cognitive Science, University of Białystok, Białystok, Poland

Editors' Introduction

The central problem discussed by the authors belonging to different disciplinary matrices is the subject of the Border. For a long time, since 2007, this international team of researchers has focused its attention on different aspects of the understanding of the border and on the impact of this understanding on different aspects of life. The border is understood as a phenomenon which not only separates but also connects different components of the natural, human, and cultural spheres of life. At the same time, the establishment of a boundary in an act of distinguishing one from the other defines the essence and objectness of that with which the human being begins to live, with which he is occupied, and which causes him anxiety. The mobility of borders creates a complex ensemble of various relations, whose essence and peculiarities the authors of this volume seek to understand.

The multinational team started its practical work by developing a joint Norwegian-Russian Master Degree programme with a philosophical focus called "Borderology: Practical knowledge experience" which is now being implemented by the academic staff of the Norwegian Nord University and the Russian Murmansk Arctic State University. A new dimension was added to this collaboration by our joint research on the border phenomenon which resulted in a series of conferences and seminars periodically held both in Murmansk and Nikel (Russia), and in Kirkenes and Svanhovd (Norway). A good example of this sustained academic cooperation is afforded by the international Kant and Bakhtin seminar which is annually hosted by Murmansk Arctic State University.[1] One way or another, these past few years have seen dozens of researchers from around the world become involved in the study of borderology, including members of the academic communities of Russia, Norway, Finland, France, Lithuania, Poland, Bulgaria, Portugal, Croatia, Hungary, and Brazil. A significant milestone in this operation was the collective monograph *Philosophy in the Border Zone (Through the Eyes of Russian and Norwegian Participants)*.[2]

[1]The sixth seminar was held in March 2018.

[2]Philosophy in the Border Zone (through the Eyes of Russian and Norwegian Participants). Edited by Viggo Rossvaer and Andrei Sergeev. Oslo, Orkana Akademisk, 2015.

This book is a result of a successful research seminar at Svanhovd research center in the border zone between Norway and Russia. The title of the seminar was *Man and Nature; philosophy and ecology in the border zone and along the Green Belt.* The Green Belt is a 12,500 km (over 7700 miles) long border zone along the one-time Iron Curtain, from the Barents Sea in North of Europe to the Black Sea in the South. This border zone is forming a corridor of habitats for an exceptional diversity of species.

With the emergence of Borderology, it became apparent that it would have to explore and clarify a large number of previously neglected questions. These were not exclusively of a theoretical interest that might have arisen from the somewhat contradictory nature of borders, fixed and mobile at the same time. Borders and border regions are particularly revealing places for social research, especially in the present era of growing globalization, and the growth of supra-state regions such as the European Union (EU). There was also an important practical aspect to these questions, since in our age of globalization individuals, ethnic groups, nations, and cultures interact with each other—and with themselves—in exceptionally complex ways, generating problems that have to be solved "here and now", "hic et nunc". Our aim is to develop a new approach to studying changes in the periphery of Europe through exploring the process in which borders themselves become visible, strengthening, meaningful, or disappearing. At the same time, we are simultaneously focusing on what those borders separate and what they bring together, along with the impact of remaking borders, which means studying the understandings of possible futures as well as the past.

Thus, the texts that we offer the reader can be seen as an attempt at reflecting on, and coming to terms with, the diverse interpretations of the border, our "point of departure" being the present time, something that enables the various conceptual and methodological tools suggested by the contributors to be applied to the very situation of our life in the border zone which both divides and connects Norway and Russia, giving a unique character to all the regional processes. This volume should therefore be read as a reference book for different approaches to the term borderology and border zone.

It is possible to identify several perspectives and thematic fields relevant to these subjects: *five parts of the book* reflect both the distinctiveness and the intrinsic interrelatedness of the individual authors' texts. *First*, the problem of the Border is considered through a discussion of the various relations between Man and Nature, proceeding from the established practices of Man's dealing with the world and with himself. *Second,* an important aspect in the understanding of this problem is revealed by the practice and theory of changing mental borders. An environmental identity can be similar to another collective identity (national or ethnic) and give us a sense of connection, possibility to be part of a larger whole, and with a recognition of similarity between ourselves and others. *Third*, another significant aspect of the understanding of the Border is manifested in the emergence and expansion of new knowledge about borders as frontiers, something that is connected with a conceptualization of new bases of Man's practical experience in his interaction with the Border. *Fourthly*, awareness of the mobility, fluidity, and change in the understanding of the phenomenon of the Border is connected with the process of the emergence of new theoretical and conceptual constructions leading to the formation of new paradigms of understanding. *Fifth*, a new interaction between the Sciences and the Humanities in their attempts to gain more knowledge about the Border is connected with a turn towards Ethics which opens an entirely new perspective on the traditional ideas of the Border and the existing theoretical constructions.

In his article "On Uses of 'Wild Nature': Empowered Versus Disempowered Agency in Kola Reindeer Herding Territories (Murmansk Region, NW Russia)", *Yulian Konstantinov* focuses his attention on the concept of "wild nature" which was created by a number of specific human attitudes developed in the context of Post-Soviet reindeer-breeding practices in the North of Russia.

Ove Jakobsen and Vivi Storsletten argue in their article "Beyond the Green Shift—Ecological Economics" that green economy, on the one hand, represents changes in the protective belt in order to defend the mainstream economy's hard core. On the other hand, ecological economics question the hard core of main stream economy.

Eduardas Spiriajevas in his article "Borderlands of Lithuania and Kaliningrad Region of Russia: Preconditions for Comparative Geographic Approach and Spatial Interaction" discusses the border zones between the territories of Lithuania and the Kaliningrad region of the Russian Federation which were traditionally perceived as peripheral zones of existence. However, in the context of cross-border cooperation such territories can come to be perceived as centers where new cultural and economic opportunities are generated, as well as new human identities. In our present-day reality, the natural is defined by human attitudes—this is the view taken by *Małgorzata Bieńkowska, Łukasz Faszcza, Łukasz Wołyniec* in their paper "Movement to Defend the Białowieża—The Problem of the Białowieża Forest Protection as an Example of a Values Conflict": one's perception of the Nature of the Białowieża Forest as either an environmental treasure to be preserved, or a natural resource to be exploited depends largely on one's own value orientations and is in many respects defined by them. Any serious attempt at comprehending the borders between Man and Nature can only be made within a special disciplinary space which is only just beginning to be granted "citizenship rights" in academic discourse: *Jan Selmer Methi* in his article "Borderology and Practical Knowledge—Humanities Response to Epigenetics?" uses the new research field epigenetics to show what borderology does on a macro scale using tools within humanities finds its parallel in natural sciences within microbiology.

Another perspective on the problem of understanding the Border is to be found in reflecting on the cultural borders of the modern European: on *Boris Sokolov*'s view expressed in his article "The Power of Doubt", the status of any border, including that of a state frontier, is defined in the present-day European culture of the era of globalization by turning to the fundamental *existentiale* of "doubt", a view that allows the author to reveal new aspects in the phenomena of migration and terror.

Basia Nikiforova in her article "Deconstruction of European Environmental Identity in the Mass Immigration Context Summary", offers to look on the European environmental identity development through rationality, growing solidarity and responsibility. An environmental rationality is not just a philosophical and theoretical enterprise, but is rooted in social practices, recovered sense of the being and formulates new arguments for mobilizing collective actions. At the same time, nowadays ethic of otherness is not a dialectic of opposites that results in the exclusion and elimination of the opposite (radical) other. In her article "Nature and Man: Crime and Punishment", *Nadezhda Golik* analyzes the ideological conception of the Enlightenment which overcame the teleological model of the understanding of Nature, making it possible for the latter to be more and more consistently perceived by Man as a means of man's existence. But in the end Nature will hit back against man's subjection. The "three senses of the word "nature", nature as wild nature, nature as dynamic activity, and nature as Stoic" highlighted in *Viggo Rossvaer's* article "Nature as Stoic", suggest another perspective on the problem of understanding the Border between Man and Nature where, in the analytic of their relationships, the context of nature remains an essential component of human consciousness.

Johan Arnt Myrstad explores, in his article "What Role Plays Intuition in Mathematics and Science? On the Borders Between Several Conceptions of What It Means to Intuit" the nuances in the use of the term "intuition" in Kant and Poincaré, thereby discovering deep paralogisms in the discussion about geometry and hence resolving some of the problems pestering modern epistemology, by focusing on the border zones of conceptions covered by the same or similar terms.

The modern European propensity for contrasting Man and Nature, sometimes placing them in direct opposition to each other, is an attitude that requires fundamental reconsideration, as *Viktor R. Tsylev* suggests in his article "Man and Nature: Approaches to the Delimitation of Concepts".

Jan Selmer Methi and Inna Ryzhkova in their article "Potential, Problems, and Challenges of Joint International Master Programmes: Case-Study of the Joint Norwegian-Russian Master Degree Programme in Borderology" reflect on the experience of academic mobility in the context of the current internationalization of education, drawing, in particular, on the experience of the long-term implementation of the joint Master Degree programme "Borderology (Practical knowledge experience)" by Nord University and Murmansk Arctic state university, and revealing new aspects in the understanding of the Border.

In his article "Human Existence Between the Forest (Nature) and Home (Technology)", *Vasilii M. Voronov* presents a project of an existential and anthropological analytic of the liminal position of the human being who has found himself caught between the regions of nature and technology. As symbolic keys for these spheres of human life, the images of the house (technology) and the woods (nature) are introduced.

Aleksandr Sautkin in his article "Cultural Images of Nature as Paradigms of Human Practices" discusses the imagination, including its archetypal attitudes, as a fundamental intermediary between abstract thinking and human nature; in this connection, the modern practices of human domination over nature and the practices of its environmental protection are considered as products of a certain attitude of the human imagination. *Andrey Vinogradov* in his article "Imagination as a

Breaker of the Border" emphasizes that borders are established in the parameters of our mind and are established by our mind; therefore, the imagination can be regarded as both the universal "violator" of borders and as their creator. It was as early as the time of mythological consciousness that the natural began to be involved in both the dialectic of the functioning of everyday practices, and in the procedures of their cognition. In *Irina Gorina's* article "Myth on the Border Between Man and Nature" the process of the symbolization of the natural world in mythological consciousness is considered in which myth becomes a means of domesticating and humanizing the world, whereas the imbalances in the interaction between Man and Nature coincide with eschatological motifs.

The functioning of established borders gives rise to a major problem, that of communication. In *Andrei Kopylov's* article "Translation in a Hermeneutical Context: Transforming Culture and Human Nature Through Ethics" communication is considered in the context of one's attitude towards the Other where the ethical aspect of understanding becomes one of the attributes of hermeneutical procedures. A multidimensional reflection on the internal borders of the human being is presented in *Andrei Sergeev's* article "The Need for Being Disinterested as a Key Characteristic of Human Nature": the author considers the need for performing disinterested actions, something that has an intrinsic ethical component, as one of the main ways of making a transition from what is contingent and particular to what is integral and one's own.

Jan Selmer Methi
Andrei Sergeev
Małgorzata Bieńkowska
Basia Nikiforova

Part I
Border Practices in the Green belt

On Uses of 'Wild Nature': Empowered Versus Disempowered Agency in Kola Reindeer Herding Territories (Murmansk Region, NW Russia)

Yulian Konstantinov

Abstract The groundwork of this paper comes from 3 months of research in the reindeer-husbandry part of the Kola Peninsula (March–May 2016). I examined the variety of uses of the 'wild nature' concept among three groups of tundra-focussed actors ('oligarchs', 'old bosses' and 'city anglers'). They fall on both sides of a 'dominating/dominated' class division in the context of post-Soviet reality. The findings suggest a reassessment of the position, inherent in critiques of 'nature-culture' dichotomies, according to which they are conceptually born exclusively for the purposes of domination. Uses of tundra space, firmly separated from an urban antipode, reveal motivations of dominance by empowered actors. At the same time, uses through which disempowered actors attain access to appealing alternative cosmologies can also be seen. Such diverse uses of tundra space I examine through their realizations in terms of tundra access and mobility. I show that a dichotomizing town-tundra ('culture-nature') division is perceived as necessary for all of them, but for different reasons. This calls for reassessing existing critiques of culture-nature bounderism. Its conceptual roots as well as instrumental motivations are in need of examination beyond current privileging of empowered actors as exclusive agents. The role of disempowered actors for sustaining culture-nature dichotomies, their diverse conceptualizations and forms of agency need also to be considered.

Keywords Bounderism · Culture-nature dichotomies · Reindeer husbandry
Kola Peninsula

Note: Unpublished or rare texts are marked PA (personal archive) and can be accessed by writing to the author at <konstantinovyulian01@gmail.com>

Y. Konstantinov (✉)
Bulgarian Society for Regional Cultural Studies, Sofia, Bulgaria
e-mail: konstantinovyulian01@gmail.com

J. S. Methi et al. (eds.), *Borderology: Cross-disciplinary Insights from the Border Zone*, Springer Geography,
https://doi.org/10.1007/978-3-319-99392-8_1

Theoretical Issues

In this paper,[1] attention is drawn to a particular problem with what can be called a 'domination bias' in critiques of boundary constructivism: namely, their privileging of empowered subjecthood. Below I clarify this point and state my reasons.

As it is well known, a principal tenet in critiques of boundary constructivism is that a theoretical tradition, anchored in the binaries of the Enlightenment and its main divide between nature and culture, is anachronistic as it imposes the seeing of variety through a single analytical lens: that of a Western (European) colonizing culture. (Cf. Bessire and Bond 2014: 1) Hence an appeal, inherent in the 'ontological turn' (i.e. Descola 2013), for addressing alterity on its own terms, thus heralding, in the words of Marshall Sahlins, 'a new anthropological dawn' (2013: xii). I would add here in parenthesis that such an appeal was made at the very dawn of American Cultural Anthropology itself. Namely, when it was realized by Franz Boas that American Indian languages and the world views of their speakers cannot be understood and should not be described in terms of the analytical categories (syntactical and morphological) inherited from Latin grammar. Neither their views of the world should be understood in the categories of the researchers' contemporary Euro-American culture. The appeal that we should look 'firmly within the bounded terrain of the Other' (Bessire and Bond 2014: 1) thus bears the mark of familiarity, which, nonetheless, does not detract from its power of persuasion.

What I find problematic, however, is a particular outcome, most categorically stated in political ecological writing of the past decades. In the writing of Donna Haraway, for instance this type of critique further extends to Western dualisms, in general, and against their main thrust of ensuring, in her words of 'One's domination of the Others: women, lower class people, non-whites, and all those whose task is to mirror the unitary self' (Harraway 1991: 177). The collective subjecthood of this 'One' bears, correspondingly, the features of 'male', 'higher class people', 'whites' and all those empowered beings that comprise the self that is mirrored. This collective subjecthood is, in other words, that of the power-holders, the empowered and the dominating. A hard and fast culture-nature boundary is seen, consequently, to place the dominant on the culture side of the binary, while the 'weak' (sensu Scott 1985) are on the nature one. Or, as it was classically phrased by Sherry Ortner in reference to gender, 'Female is to male, as nature is to culture' (Ortner 1974).

Within what I have called above the 'domination bias' in critiques of boundary constructivism, a counter, deconstructive thrust follows the basic tenets of the ontological turn. Namely, by claiming that in the real world the binary of culture as different from nature does not exist, it is proposed that what does exist—on the

[1]Field-research was supported by 'Humans and the tundra nature of the Kola North: interactive social practices', Research Project No. 17-13-51601, supported by the Government of Murmansk Region and the Russian State Science Fund (*RGNF*) in 2017, following the results of the Regional Competition 'The Russian North: History, Modernity, Prospects'.

strength of ethnographies in non-Western settings—is a single all-encompassing culture, revealed in an infinite multitude of forms. This insight, traceable in its literary projections at least back to Marcel Proust's writings of the first decades of the twentieth century. [2], found a reincarnation in the perspectivism of Viveiros de Castro (2000) and his followers.

From this point on, the very existence of hard and fast dividing lines and binarism itself became, as it were, politically incorrect. In postmodernist epistemologies, a virtual abhorrence of hard and fast dividing lines came to be noticed, with a corresponding preference for the amorphous, fluid, fuzzy-edged and ambiguous. An intellectual culture and aesthetic of fluidity or liquidity (Bauman 2000) came to replace one of firmness.

In this paper, I do not address the issue of whether the relationship between 'culture' and 'nature' can be defined as a binary or not. In other words, whether it is right or wrong to see them as entities, separated by a hard and fast boundary, or, alternatively, to see them as a unity ('culture') with an infinite number of embodiments, each with its own instantiation ('perspective') of a common universe. I would note here, that in the second case, one subject of a particular perspective is not sharply different from another—like human from animal, for instance but is more or less like them since they belong to a common, as it were, 'deep structure' (sensu early Chomsky (1957)).[3] This 'more or less' makes dividing lines between them finely graduated or 'fuzzy'[4], and they themselves tend to become rhyzomatically connected, rather than standing as members of binaries.

The task which I set before myself is much more modest. I am questioning the premise that motivations and even applications of 'firm boundering' have, exclusively, domination as a target, and those in power—as subjects. My argument is that while we can find abundant proof of social constructivism with this particular aim in mind, we can, at the same time, find also numerous instances in which disempowered agents make use of and actively resist the erosion of boundaries. A dissolution of boundaries may be experienced as the advance of chaos and, closer to my field-case 'lawlessness' (Rus.: *bespredel*). Importantly, it may be resisted on the grounds that it is an obstacle to a search for alternative realities by disempowered actors—sought in attempts to escape from or resist domination. In this way, the defence of a 'culture/nature', '*domus/agrios*' firm boundary may unexpectedly appear as an important item in the 'weapons of the weak' arsenal (for the felicitous phrasing James Scott 1985). If so, the connection between boundary constructivism—in my case 'culture/nature'—and its political implications (as an

[2]Literary premonitions of perspectivism can be seen in the works of French writer Marcel Proust. In the 1920s, he wrote: 'The only true voyage of discovery, (…), would be not to visit strange lands but to possess other eyes, to behold the universe through the eyes of another, of a hundred others, to behold the hundred universes that each of them beholds, that each of them is (…)'. (Proust 2006: Vol. II, 657).

[3]For a critique of this view Bessire and Bond (2014).

[4]For the origins of the concept, see Zadeh (1965) on 'fuzzy sets', Lakoff (1973) on 'fuzzy concepts'.

exclusive 'weapon of the strong') becomes problematic in its subjecthood part. A demand or at least an active search for territoriality (sensu Sack 1986) effected through firm boundering, may in other words, come as much from 'below' as from 'above'.

In this article, I draw on empirical evidence from the reindeer-herding part of the Kola Peninsula to defend such a 'bounderism from below' argument. The evidence comes from longitudinal ethnographic work with reindeer herders from Lovozero District since the spring of 1995, with a latest focussed research—leading up to this paper—conducted in February–June 2016.

The Setting

This is the grazing range of Reindeer Herding Cooperative '*Tundra*' of Lovozero, specifically: that of its Reindeer Herding Team No X.[5] The territory stretches from the Keivi Ridge in the SW (winter pastures) to the Barents Sea Coast to the NE (summer pastures). It is controlled by the Lovozero Village Administration and rented out (for a symbolic annual fee) to the Cooperative. Apart from this inheritor of the former Soviet Farm (*sovkhoz*) bearing the same name, at present numerous other leaseholders make use of the territory. They are mainly owners of salmon fishing and/or hunting camps.[6]

The focal point in my setting is one such camp. It was created on the site of a closed down Hydro-Meteorological Station, situated in the central part of the grazing range. Very soon after the closing down of the Station in 2010, it was leased for a period of 25 years to a businessperson from a regional town.

The Station, now private hunting and fishing lodge, is situated on the top of a hill overlooking a big lake, with the buildings of Reindeer Herding Camp X, at its foot. Camp X or Base X for short, to approximate local usage, is the second focal point, the third being my own permanent camp, some two kilometres to the south of Base X. Let me add here that since June 2016, the camp became officially a Research Station of Murmansk Arctic University.

The Actors

I have divided them into three groups: 'oligarchs', 'old bosses' and 'city anglers', with the current generation of reindeer herders as background actors.

---

[5]Whenever possible I have changed designations, place name, and personal names in an effort to preserve a degree of anonymity of my informants.

[6]On such camps Konstantinov (2016), Ogarkova (2007), Osherenko (1998).

The **'oligarchs'**, representing new power over the territory, had as a central figure Andrey,[7] the businessman renting the former Meteorological Station. This for a symbolic sum, as the rumour went among the herders. The takeover, as tundra rumour also had it, had been achieved thanks to Andrey's connections with influential persons in the regional administration, as well as with the help of a high-orbit 'real oligarch' from Moscow. The new owner and his rich and influential friends, who came to the Station for fishing and hunting trips were generally dubbed 'the oligarchs' by the reindeer herders. I follow this usage here.

The **'old bosses'** were the former heads of the Station and Reindeer Herding Team X, respectively. They were both old-age pensioners[8] by the time of the study and had come to fish for a month as guests of the current reindeer-herding team. Naturally, both of them—Dad Vasya and Uncle Kolya as the younger reindeer herders would respectfully address them—deeply resented their fall in status. Still, some sort of peaceful coexistence was maintained between the 'bottom' (the reindeer-herding camp) and the 'top' (the 'oligarchs' lodge) of the hill.

The third group was of two young **anglers** from Murmansk: Ivan (Vanya), an engineer, working in the Fishing Port of Murmansk, and his friend and colleague from the Port—Gennadiy (Gena). Vanya and Gena were both in their mid-30s, passionate anglers, spending practically all their weekends and other stretches of spare time on angling outings near Murmansk. This was their first time so deep into the tundra. 'It is only recently I could finally buy the (Yamaha) 'Viking' (snowmobile) so we could get here', Vanya explained when the topic was broached, 'and Gena still cannot afford to get one. We are not like those up there (*na verhu*) (pointing with his hand in the general direction of the Station) with all the machinery and petrol up to here (a gesture across his throat)'.

Control Over Resources: The Vassiliy-Ty Incident

This event, which happened soon after the arrival of Vanya and Gena, illustrated who had control over the territory in the new times. Although the city anglers were not affected by it themselves, it was their descriptions and discussions of it that forcibly related an important message. Namely, that they resisted intrusion into the tundra by people like the 'oligarchs' or other potential leaseholders. This resistance was for the sake of preserving the tundra in its 'wild' state and thus allowing its use as a way to access reality sharply different from their daily urban one. I describe the case in some detail.

Vassiliy-Ty is a lake in the NE part of the grazing range of 'Tundra', in the spring–autumn pastures of the reindeer herding team in my story. During the heyday of Soviet reindeer husbandry and until late after its demise, reindeer herders

---

[7]All personal names are fictitious: see Footnote 5.

[8]Note that the current pensioning-off age in reindeer husbandry is 50 years.

considered it their own prized fishing location. Many parts of the herding calendar were organized in view of fishing there.

Following these established ways, the two former bosses, Dad Vasya and Uncle Kolya, drove by snowmobile to the lake, but just as they were about to set their nets under the ice, two private guards—*chopovtsy*[9]—suddenly swooped on them on some rather fancy-looking snowmobiles. Didn't they know the lake was off-limits without a 'special arrangement'?

Kolya and Vasya did not know the guards themselves, but the manager of the salmon fishing company, which had exclusive rights of this part of the territory, was known to Kolya from former Soviet days. This was used as an instrument to bear on the situation. Consulting between themselves a little, the guards announced the following decision: for old times' sake, let Kolya and Vasya set their nets and fish the lake, but on the eastern side only and for not more than three days. If they caught them on the lake after that, the conversation would be 'difficult', as they put it.

It so happened, that during the three days allowed, two track vehicles (*vezde-khodi*) of the Cooperative had arrived at the Base, there was much celebrating (i.e. drinking) at the huts, as brigade traditions required it, and in result, Kolya and Vasya were fit enough to see to their nets only on the fourth day. But when they got there, the nets had been already gathered in and taken away. The guards had been true to their word: three days, but no more.

While the former bosses were having their troubles at Vassiliy-Ty, the 'oli-garchs' had gone out on an outing for salmon at the much prized and well guarded Vostochnaya Litsa Lake. A few days after that, they stopped by my camp, where the city anglers happened also to be. 'How was it at the Litsa?', they asked, 'were there any guards prowling?'. 'Oh, yes, they came, sure enough', Andrey answered. 'And what?' 'We reached an agreement' was the answer.

The expression 'to reach an agreement' ('*my dogovorilis*') is worthy of attention here. In all probability, what it meant was that it was made known to the lake guards in question, who the oligarchs were and, in particular, with what people they were connected at the higher rungs of the regional administration. It also contains the hint of a bribe given, as when, for instance a controlling organ—like road police—would find fault with a driver. We all three discussed it afterwards together with the former bosses when they came to see us. They were of the opinion that it was enough to mention the oligarch from Moscow to make the guards change their tone.

[9]From *ChOP*, acronym for *Chastnoe Okhranitel'noe Predpriyatie* (Private Protection Enterprise).

Mobility: The Helicopter

The second profile event was the coming of a helicopter: a big Mi-8 machine of a private company. The aircraft landed at the helicopter pad of the Station, which was part of its older infrastructure. Since the Station used to have the status of 'difficult to reach' (*trudnodostupnaya*), while it existed, it was often that helicopters landed on that pad: either to bring people or goods to the Station, or in connection with the needs of the reindeer herding camp below. The herders remembered nostalgically that helicopters in those days, when the price of petrol was of little concern, would fly back and forth even on 'small errands' (*po melocham*). Examples would be quoted of helicopters bringing in parts of personal belongings left behind in the village. 'The coal for heating the Station would come in by helicopter', Dad Vasya remembered during this particular conversation. 'Small things also: the pilots were glad for an excuse to do some trout-fishing or get fish or (reindeer) meat from us'.

For the young anglers from town, all of this was new. They had grown up during the time when petrol had already become a serious concern, as it was for them right at that moment: they had estimated that on the petrol they had, they could hardly make it back to the village. The contrast with the previous liberal times was painfully accentuated. 'So these lords (*bare*) up there, they can summon helicopters just like that? To bring them spares?' It had become known, by that time, that the helicopter had been summoned to bring a new gearbox for the oligarchs' ATV. Uncle Kolya had learned it from Ivan Ivanich: one of Andrey's servants. 'Yes, like that' Uncle Kolya flicked his thumb against his forefinger. 'They drive all over the tundra also: they have the petrol and also the connections'.

Metaphors

Petrol and connections summed up metonymically the new order in the tundra. They stood for wealth and power. The group of the 'lords' above represented the new order in their tightly knit group of business people ('oligarchs') and administrators (state employees, *chinovniki*). For various different reasons, all of those 'below', i.e. at the Base by the bank of the Lake and at my camp, had found themselves to be outsiders in a newly structured tundra universe, or even trespassers, as the Vassiliy-Ty incident illustrated. For the former tundra bosses and the reindeer herders in general this was a very sore point. During the previous order, they had been publicly celebrated as 'custodians of the tundra' (*khozyaeva tundry*). It was, as it were, their native element and a territory under their control. For the young anglers, those were days dimly remembered, but they also felt themselves deprived of being barred from engaging in pursuits of an encounter with 'wild nature'. As I illustrate shortly, the newly powerful expressed and acted out their motivations in a similar way. The motif of 'othering nature' could be seen, therefore, to override the new class divisions.

The coming and going of the helicopter did not merit much discussion at the Base. It was remembered that Volodya, the big shot from Moscow and Andrey's friend and patron, had once arrived in his own helicopter, as it befitted an oligarch of his rank. The immediately pragmatic side of the matter was, however, what oriented conversations: the fact the Mi-8 had brought in a considerable amount of petrol. Among the constantly petrol hungry herding community, this was a matter of utmost attention: it meant the ability to move both for contacting herd fragments, strewn over the tundra, and fishing in trout and salmon spots, closer to the Barents Coast. Using Uncle Kolya's connections with the oligarchs' servants, particularly with Ivan Ivanich, a plea was made for getting three hundred litres on credit. This was granted by the oligarchs at 50 roubles the litre.[10] Speedy preparations for the trip began. To my questions about how they would tackle the various private guards who may intercept them, they proffered a valid reason. This was the herding territory of the Brigade and they had all the right to do their herding business there.

The helicopter affair and its ramifications impressed the young urban anglers. How the tundra was parcelled out to various private interests they knew in principle, but here they were looking at it directly, day by day. The summing up of this new experience was done by Vanya. While we were discussing the possibility for them to ask the oligarchs for 50 litres of petrol to get back to Lovozero, he said: 'You see, we can't be like them, it is too late in the day now, and we wouldn't have done it anyway. For us fishing means something else, another reality'.

This was a crucial statement in my view. What Vanya was saying was that they would not have not chosen the ways of vertical mobility of the 'oligarchs', i.e. of rising from petty trading to vegetable wholesaling (as was Andrey's case), or becoming pig-feed kings (as that of the Moscow oligarch). Or more precisely, as they had heard it from the herders: of using connections from former Komsomol days to buy at Soviet prices on credit, and sell at the new liberated prices after the reforms of early 1992. These ways to make the millions were memories of the past now, but for them they were 'not interesting', as they put it, meaning they were morally untouchable for them. 'For us the tundra is a chance to get out of such things, it is a parallel reality. We can still find spots in it where these types have not reached, but the spots are getting fewer every day'.

It is to be noted at this point that in all such conversations and with all groups concerned—the urban anglers, the herders, the former bosses, and finally: the oligarchs—'tundra' stood for 'nature', or rather: for 'wild nature'. In the conversations with all four groups, the use of the term implied distancing oneself from a referentially critical locus, and in this initially geographical way, establishing an opposition. The themes differed and, consequently, the terms that captured the opposite end of the dichotomies. For the herders, it would be 'the settlement' (*poselyok*), from which they commuted to the 'tundra' for performing various official and private tasks, intricately mixed with one another. This theme paralleled

[10]The average price of petrol (A 95) was RBL 33/l. at this time, so the oligarchs did not raise the price excessively.

the one of the OAPs, with the additional nuances of their case. Namely, that the settlement was not so attractive anymore for them—as it was for the much younger, working herders—while the Base by the Lake and the surrounding tundra territory was the place where they had spent their lives. They wished if not to return to live permanently there, at least to have the opportunity to be there at exciting fishing or berry-picking times.

In the case of the oligarchs, the opposite pole was described in terms of the enormous stress that a crushing load of work big business imposed. The self-representation was of hard-working people. They had to overcome the negative effects of the present period of economic crisis, when people had less money to spend. Another major obstacle, according to them, was the mentality of employees. 'These are the same old soviet beings (*sovki*) of before.[11] 'Andrey would complain. 'I am sick and tired of repeating that I am not a Soviet factory from which you can pilfer as much as you like and pretend to work. It is a private business, in which I have invested my life. The tundra is the only place where I can get away for a while from tension and stress. Some people go abroad, to Egypt, Spain, such places. I prefer to come here with my friends. Here we charge our batteries! (*zdes' zaryazhaemsya*)'.

The world of the city and business activities was thus described as stressful and one from which escape routes were being sought, ones leading, in the particular case, to the tundra and its opportunities for engaging with 'real things'(*nastoyashchye veshchi*), ones that befitted real men. 'I often take my son on such trips', Andrey went on 'let him learn to be a man'.

In this particular narrative, the tundra was represented as a 'get away' place, an alternative reality with its particular attractions. City-life or 'civilization' was portrayed as a place or reality inhabited by necessity, as it were, the attractive place was here, in the tundra. In a constellation of possible 'worlds', or frames of reference, the centre of gravity in terms of attractions lay here. The latter came out as ones offered by the kudos of blood-sports: of shooting a spectacular animal like a bear or moose, or of catching some trophy-calibre salmon, trout or pike.

The poetic trope of representing the 'wild' as the place of manly virtues, with the city as the place of vice, takes us back to the very beginning of what Andrey referred to as 'civilization': to the *Gilgamesh* epic, or much later to Tacitus' *Germania*, through the 'noble savage' sentimentalism of the 18th c., the romanticism of the early nineteenth and the Indianomania of the late. And so on, in various representations, reaching as alive as ever to the present day. Among the multitude of ideological strands that the contemporary situation I describe leads up to—like its strongly gendered, 'manly' emphasis—I single out here the similarity, and yet critical difference as it concerned the 'escape ideology' of the young anglers, as well as, in variations of a 'disempowered' narrative, of the other actors: the herders, and the 'old bosses'.

[11]*Sovki* (sg. *sovok*) is a Russian pejorative colloquialism for 'Soviet beings', i.e. people with die-hard Soviet mentality, unable to imagine life in other than Soviet terms.

In the young anglers' narrative, the escape route to the tundra was motivated by very different reasons compared to those of the oligarchs, retaining the similarity of the 'city' as a place of necessity from which one needs to get away at the least opportunity. The former Soviet reality was positively represented in the youngsters' narrative: as a fairer and more virtuous reality, strongly contrasting it with the present one of lawlessness and greed, in which 'those like our friends at the top', (a gesture towards the Station), 'do as they please'. The incident with them being able to 'reach an agreement' with the private guards at the Vostochnaya Litsa Lake would be quoted (in contrast with the confiscation of the pensioners' nets), as also the unlimited quantity of petrol at their disposal, their glittering transport machinery, their servants, their summoning of a big transport helicopter at the flick of the fingers. 'We are trying to get away from all that', Vasya summed it up 'but, as you see, we are here living next door to them and have to be nice, so they sell us some petrol'.

A parallel reality thus seemed possible only in case one managed to get to some 'really wild' place, beyond the reach of current power-holders. This topic: of where 'really wild places' were would be discussed with the herders and the pensioners, often only to establish the fact that a salmon-fishing lodge for rich foreign and home clients had already been built there and the surrounding territory was already off-limits.[12] Still, places would be named, in between the stretches of tundra already appropriated.[13] Fabulous places, where birds would alight on your head and animals look at you in wonder, as they had never seen a human being before.

Whether such places still exist on the Peninsula, or can be reached by some next to magical means is a matter of speculation, or, more precisely: of dreams. The important issue is their need to be as a realm of escape. The two city anglers, together with all others at the Base (including myself), wished to believe that 'wilderness' existed and we can reach it. Plans were made and treks discussed. The vezdekhods had to start back before the tundra became impassable with the spring thaws. The young anglers' 2 weeks of leave was reaching its end. The tundra expedition finished on a hopeful note: the search for the 'really wild' stretches would continue.

Conclusions

I have tried to show in this paper how in a post-Soviet northern context, 'boundering'—that is, the drawing of a dividing line between 'wild nature' (tundra in my case)—and the spaces that stand in opposition to it, follow desires of those currently in power, as well as from those of the currently disempowered. Those 'at the top' (in my case: at the former Station) reiterate the script of an age-long political

[12]On high price-range salmon-fishing camps see Note 6.

[13]On similar 'intersticiality' in a Siberian context Anderson (2006).

ecology. In it, the exclusive access to 'wilderness'—be it where in Medieval times the 'King's deer' roamed (cf. Eliason 2012), or when it encompassed the Victorian countryside as a whole, before the Game Act of 1831 (Fisher 2000; Archer 1999), served as an exclusive playground, in which blood-sports provided a diversion from duties or boredom, and an opportunity for the honing of 'manly' skills. It simultaneously effected and legitimized a superior social position. In my field-case, such Medieval and Early Victorian tropes—not to mention those of the rich Russian/Soviet heritage—were reproduced as exclusive access to salmon fishing and tundra big game hunting.

'Wild nature' can be, at the same time, the place of escape from such an order, where power arrangements are renegotiated on the risky terms of the trespasser. This line connects the investigation with poaching practices, particularly such that have political motives: either for the preservation of age-long cultural values and rights, or as rebellious acts against the dominant classes and their legislature.[14] This strand needs examination as regards its post-Soviet representations, which has so far been lacking.

Another important issue, which to my knowledge has not been addressed so far as regards 'boundering from below', is its pronounced gender aspects, namely that it is a territory, in the Russian case at least, overwhelmingly populated by men. When we get to the related issue of 'political' poaching, the absence of women becomes even more striking. While in fishing and hunting women, in many cases around the globe may have a strong presence (i.e. Sami women in fishing, Evenki women in hunting, etc.), poaching is a domain, where women seem to be next thing to absent, or, at least, we do not hear of the opposite. This gender specifics may be related, or indeed may stem from the fact of 'female flight' from places, perceived as remote and 'wild'. It may also be, on a more general level, that escape and seeking of alternative frames of reference ('worlds') is a gender-specific issue.

Investigations here are at their beginning and it has borne the intellectual stamp of women's studies. This aspect of historical development of the investigation of 'boundering' has possibly led in its turn to discussing nature/culture boundaries as an exclusively empowered men's construction for purposes of domination. I have tried to show here, that in view of the existence of cases such that I have presented above, constructing this boundary may be seen also as the 'work' of the dominated, which problematizes, or at least serves to expand existing views.

References

Anderson DG (2006) Is Siberian reindeer herding in crisis? Living with reindeer fifteen years after the end of state socialism. Nomadic Peoples NS 10(2):87–104

Archer JE (1999) Poaching gangs and violence: the urban-rural divide in nineteenth century Lancashire. Brit J Criminol 39(1):25–38

[14]For general typologies, see von Essen et al. (2014); Eliason (2008, 2012).

Bauman Z (2000) Liquid modernity. Polity Press, Cambridge
Bessire L, Bond D (2014) Ontological anthropology and the deferral of critique. Am Ethnologist 41(3):440–456
Chomsky NA (1957) Syntactic structures. Mouton, The Hague
Descola P (2013) Beyond nature and culture. University of Chicago Press, Chicago
Eliason S (2012) From the King's deer to a capitalist commodity: a social historical analysis of the poaching law. Int J Comp Appl Crim Justice 36(2):133–148
Eliason S (2008 [1999]) The illegal taking of wildlife: toward a theoretical understanding of poaching. Hum Dimensions Wildl 4(2):27–39
von Essen E, Hansen D, Nordström Källström H, Peterson MN, Peterson TR (2014) Deconstructing the poaching phenomenon. A Review of Typologies for Understanding Illegal Hunting. Brit J Criminol 54:632–651
Fisher J (2000) Property rights in pheasants: landlords, farmers and the game laws, 1860–80. Rural Hist 11(2):165–180
Harraway D (1991) Simians, cyborgs, and women: the reinvention of nature. Routledge, New York
Konstantinov Y (2016) Tourist hegemonies of outside powers: the case of salmon-fishing safari-camps in territories of traditional land-use (Kola Peninsula). In: Viken A, Müller D (eds) Arctic indigenous tourism. Channel View Publications. (In press)
Lakoff G (1973) Hedges: a study in meaning criteria and the logic of fuzzy concepts. J Philos Logic 2:458–508
Ogarkova MN (2007) Organisatsiya sportivnogo rybolova na territorii Murmanskoy oblasti (Na primere ZAO "Reka Ponoy") [Organization of sport fishing on the territory of Murmansk Region (The case of Shut Joint-stock Company 'River Ponoi')] Diploma paper, Vologda: Vologodskiy gosudarstvennyy tekhnicheskiy universitet, [Unpublished MS, PA]
Ortner SB (1974) Is female to male as nature is to culture? In: Rosaldo MZ, Lamphere L (eds) Woman, culture, and society. Stanford University, Stanford, pp 67–89
Osherenko G (1998) The Ponoi River report: sport fishing in the Kola Peninsula. A Project of the Circumpolar Conservation Union (Unpublished MS, PA)
Proust M (2006 [1927]) Remembrance of things past, vol II. (Translated by Scott Moncrieff CK and Hudson S). Wordsworth, London
Sack RD (1986) Human territoriality: its theory and history. Cambridge Studies in Historical Geography 7. CUP, Cambridge
Sahlins M (2013) Foreword. In: Beyond nature and culture. Philippe Descola. University of Chicago Press, Chicago, pp xi–xiv
Scott JC (1985) Weapons of the weak. Everyday forms of peasant resistance. Yale University, New Haven and London
Viveiros De Castro E (2000) Cosmological deixis and Amerindian perspectivism. J R Anthropol Inst (N.S.) 4:469–88
Zadeh LA (1965) Fuzzy sets. Inf Control 8:338–353

Borderlands of Lithuania and Kaliningrad Region of Russia: Preconditions for Comparative Geographic Approach and Spatial Interaction

Eduardas Spiriajevas

Abstract The focus of research based on geographic comparison of borderlands of Lithuania, and Kaliningrad region of Russia, with special aspect on the Curonian lagoon coastal-border region, which is located between two countries along the Baltic Sea. The border areas of these neighboring countries have common and different historical development. These borderlands researched in historical, cultural perspectives, and nowadays there are being conducted significant researches of different territorial units in the forms of cross-border or trans-border regions, the groups of wards or municipalities as well. Nevertheless, in the frames of regional and national policies, these territories recognized as peripheral ones, but in the strategic point of view, these territories are important for elaboration of joint actions and new identities. Thus, the borderlands researched in an approach for comparative aspects in order to reveal more detailed understanding and definition of their role in cross-border and trans-border activities, social-economic development and spatial interaction. Relatively, the lack of geographic comparative knowledge about borderlands of Lithuania and Kaliningrad region of Russia, that makes these areas more peripheral in regional context and adjacent for applied geographic and comparative researches. Tourism as a process has a key role of increasing an attractiveness of cross-border regions in order to renew the physical and cultural importance of border regions and to maintain their spatial interaction for joint actions.

Keywords Borderlands · Lithuania · Russia · Spatial interaction
Development

E. Spiriajevas (✉)
Klaipeda University, Klaipeda, Lithuania
e-mail: borderlithuania@gmail.com

J. S. Methi et al. (eds.), *Borderology: Cross-disciplinary Insights from the Border Zone*, Springer Geography,
https://doi.org/10.1007/978-3-319-99392-8_2

Introduction

The most intensive scientific and applied researches of Lithuania's borderlands and border areas with Russia, Belarus, Latvia, and Poland appeared in 1960s–1970s. Nonetheless, the first reliable data on borderlands comparative researches appeared and systematized in 1953–1955, and since this period, detailed physical and economic comparative researches of borderlands have been conducted. The topics of these researches based on analysis of structure of human resources, their density and distribution, on structure and functions of settlements as well. In addition, an important issue for analysis was the structure of land-uses, natural resources like deposits of drinking water, sand, clay, etc. The outcomes of these researches have been applicated for planning of economies in agriculture and industry for efficient allocation of work force in border and peripheral areas. Historically, there are many indicative facts, which point the use of comparative analysis for rural-border areas. These researches have been applied for maintenance of better functioning of the former collective farming in the borderlands close to neighboring at that time former soviet republics like Latvia, Belarus, and the Kaliningrad oblast of Russia, which is an exclave today in particular.

After the collapse of the Soviet Union, when Lithuania regained its independence, the comparative researches more in detail started to implement in the analysis of the newly delineated borderlands. These topics were on the problematics of land-uses, ecological-political challenges, dynamics and structure of human resources in border-rural areas, localization of natural resources in borderlands. All these topics became a focus for international political discussions, negotiations and important prospects for the joint international agreements.

The following Lithuanian geographers and researchers on regional studies have conducted the significant geographic comparative researches of Lithuanian borderlands: Liekis (1997), Baubinas and Stanaitis (2001, 2002a, b), Baubinas et al. (2002), Burneika (2003), Stanaitis (1996), Vaitekunas et al. (2001), Spiriajevas (2004a, b), Buciene and Spiriajevas (2005), Burkšiene and Spiriajevas (2015).

Nevertheless, until now there is a strong need for reliable statistical data and actual information for performance of social-economic analysis of borderlands by different comparative geographical criteria and aspects. Thus, the conducted comparative analysis of neighboring border areas carries out valuable results for better understanding, determination and significance of borders and borderlands of neighboring countries. The participation in international programs, which is granted by the EU, elaboration of joint actions in the field of municipal and entrepreneurial cross-border and trans-border cooperation, these processes generate the need for analysis more in detailed of borderlands (areas, territories, and regions).

Recently, due to increased geopolitical and political tensions between the EU's and Russia's political bodies, geographic researches of Lithuanian—Russian borderlands are not on a wide scale. The main comparative geographic researches are encompassing the coastal areas of the Curonian lagoon, as well on development of tourism infrastructure and recreational activities, conditions for social-economic development according to various criteria and indicators. A majority of these researches are conducting by research institutions. The outcomes of these researches based on expertise knowledge. A social-economic estimation of borderlands quite often does not correspond to the official opinion of public authorities, and it can be argumented differently by researchers.

According to different sources it has been determined that on Lithuania's side there were conducted a lot of significant researches related to the issues of ecological problems, land-use, and sustainable development, as well on perspectives related to recreation and tourism development. In the meantime, on Russia's side there were conducted researches, which have a vivid economic applied character, impact of exclaveness on Kaliningrad region's cross-border and trans-border activities, cooperation, and economic development in general.

In Russia, there are conducted scientifically meaningful comparative researches of borderlands between Russia-Ukraine, Russia-Finland, Russia-Estonia, Russia-Kazakhstan, Russia-China. Concerning the Kaliningrad region of Russia, there are conducted more comparative researches on borderlands with Poland as well and other related researches in the context of positioning the Kaliningrad region within the Baltic Sea region. An analysis of Lithuanian-Russian borderlands still has a lack of scientific based evidences, in comparison to the Polish-Russian borderlands.

Significant geographic comparative researches conducted by the following Russian economic geographers: Fedorov and Korneevets (2015), Bilchak et al. (2000), Pustovgarov (2002), Bolychev et al. (2015), Druzhinin (2016), Klemeshev (2004). The Polish economic geographers as Palmowski et al. (2003), Palmowski (2004), Kulesza and Maciejewski (2003), have conducted the significant geographic comparisons of Polish-Russian border. The topicality of borders and borderlands and their regional policy also have been researched by Paasi (1996), Timothy (2001), Szlachta (2004). Some important geographic aspects described in the following publications as in Euroregions … (2002), Introduction … (2002).

The purpose of research is to accentuate the topicality of comparative geographic approach on a problem of spatial interaction in the borderlands of Lithuania and the Kaliningrad region of Russia, the main focus concentrating on the Curonian lagoon region.

The tasks

- To maintain geographic approach on the concepts related to borders, borderlands and activities;
- To compose the set of criteria for implementation of geographic comparative research based on example of borderlands between Lithuania and the Kaliningrad region of Russia.

- To represent a role of tourism (natural and cultural) for increasing the spatial interaction in the Curonian lagoon cross-border and coastal region.

Generally, a plenty of different comparative researches of borderlands and border areas have been conducted. Thus, the author presents only the main guidelines for comparative geographic approach in order to promote the spatial interaction between the borderlands of Lithuania and Russia.

Preconditions for Comparative Geographic Approach

In the studies of borders and borderlands, there are mentioned different and various definitions on the following concepts like *border*, *borderland*, *border area*, *border region*, *cross-border*, *trans-border*. There is a need to elaborate the common definitions of these basic concepts with geographic explanations, what is essential for continuation of the studies on borderology. The key point for geographic comparison of borderlands suggested to be based on the following geographic definitions (Table 1).

Particularly, the issues on comparative geography of borderlands are obviously seen in examining of satellite images that vizualize geographic differences,

Table 1 Geographic definitions of basic concepts related to studies on borderology

Concepts	Definitions
Border	Real or imaginary line that determines the edges of certain subject or objects, and eliminates the subject or subject from others
Borderland	Territorial system that is under the impact of border factors
Border strip	Territory wherein begins 5 km distance before state border
Border cordon under strict control	Territory within the distance from 1 km to 50 m before state border
Border zone	Territory wherein begins 15 km distance before state border
Border area (territory)	Administrative territorial units that are nearby state border
Border region	Territorial system that is under the impact of the group of factors related to borderland
Cross-border cooperation	General or joint activities that are being implemented by subjects or objects on the both sides of border in cross-border regions
Trans-border cooperation	General or joint activities that are being implemented by subjects or objects in their countries and regions that are under the impact of factors of border and borderland
Spatial interaction	The movement of people, goods and ideas within and across the geographic space

Author's own research material, 2016

Fig. 1 Comparison of Land-use Structure in Borderlands (The Nesterov district of the Kaliningrad region on the left, and the Vilkaviskis district of Lithuania on the right). *Source* Satellite image by Maps of Google, 2017

similarities, and spatial structural changes. It is an essential methodological tool which applicates the retrospective and retrogressive analytical approaches in comparative analysis of landscapes. In Fig. 1. represented different structure of land-uses in borderlands in the territorial range, wherein the fertility of soil is rather higher than in other regions. This geographic image indicates the role of border factor in suppressing the spatial interaction between two states in developed systems of agriculture that based on traditional functioning of land-uses. Figure. 2 visualizes the Nemunas river as the natural border between two states. The landscape has the same genesis. Due to changed traditions of farming, the land-uses are not in intensive use anymore. Some remaining as old polders indicate the former intensity of agriculture in border zone, but recently this territory is a peripheral in both countries.

In Fig. 3 is visualized, evoluted different landscapes that are delineated by the natural border of the Nemunas river in the Rusne island. Lithuanian part is more reclamated, populated while Russian part is unpopulated, and has a very low degree of landscape's reclamation for economic needs. On Lithuania's part there is the Nemunas river delta regional park, and on Russia's part, the nature reserve of the Neman river delta. The juridical status of these two protected territories is different and either local or regional authorities cannot sustain any joint activities in order to start cross-border cooperation or to elaborate the activities related to promotion of spatial interaction in this border zone.

The borderlands between Lithuania and Russia are in extensive natural and social use, except the areas in the Curonian spit, where the processes of

Fig. 2 Comparison of Land-use Structure in Borderlands along the Nemunas River (the Pagegiai municipality of Lithuania above, and the Neman district of Russia belove). *Source* Satellite image by Maps of Google, 2017

Fig. 3 Comparison of Land-use Structure in Borderland in the Nemunas River Delta (the Silute district of Lithuania above, and the Slavsk district of Russia belove). *Source* Satellite image by Maps of Google, 2017

international tourism are in a rather high intensity on the both sides of the border in the spit. Due to a high attractiveness of natural and cultural recreational resources in the spit, these border areas are recognized as priority areas for international tourism development and for increasing of spatial interaction. Due to a geopolitical tension, the spatial interaction in the field of promotion of tourism and recreation is limited and not spatially and efficiently integrated on the both sides of the Curonian spit and the Curonian lagoon as well. A boundary in this region rather suppresses than stimulates the reciprocal actions for spatial interaction.

Criteria for Comparative Geographic Analysis in Borderlands of Lithuania and in the Kaliningrad Region

An application of geographic approach for comparative analysis requires elaboration of a certain order of researches. In the previous researches, it has been determined that analysis of borderlands is various in topics and topicality, but still there is not conformed any systematic methodology or at least approach for researches. Thus, the researchers are actual, but weakly interrelated by different and logical criteria. There are presented recommendable the following types of criteria for conducting of comparative geographic analysis of borderlands between Lithuania and the Kaliningrad region of Russia:

- Geographic;
- Historical;
- Political;
- Social;
- Economic;
- Cultural;
- Infrastructural;
- Touristic.

These types of criteria have been applied for conduction of comparative analysis of borderlands together with retrospective and retrogressive methodological approaches. *Retrospective*—is an analysis of the past for figuring out the processes and consequences in nowadays, and *retrogressive*—is an analysis of nowadays for figuring out the processes and consequences of the past. The recommendable types of criteria and criteria for comparative analysis of borderlands presented in Table 2.

In the research revealed the following comparative physical and social geographic outcomes, that:

- a length of Lithuanian–Russian borderline is 290 km in total, and 249.3 km of this distance is the land and river border. Therein, 18.5 km a water border within the Curonian lagoon, and 22.2 km a water border with the Baltic Sea;

Table 2 Comparative criteria of borderlands: Lithuania and the Kalinigrad Region of Russia

Types of criteria	Criteria		
1. Geographic	1.1. Physical	1.2. Social	1.3. Common
2. Historical	2.1. Period until 1945	2.2. Period from 1945 until 1990	2.3. Period since 1990
3. Political	3.1. Geostrategic	3.2. Territorial administrative division	3.3. Cross-border and trans-border cooperation
4. Social	4.1. Demographic	4.2. Population and distribution of settlements	4.3. Well-being
5. Economic	5.1. Sectorial	5.2. Investments	5.3. Incomes
6. Cultural	6.1. Ethnical and religious composition of population	6.2. Cultural heritage	6.3. Traditions and behavior
7. Infrastructural	7.1. Technical	7.2. Social	7.3. Economic
8. Touristic	8.1. Economic	8.2. Cultural	8.3. International

Author's own research, 2017

- about 85.96% of the land and river border lasts along the Nemunas river, the Sesupe river, the Vistytis lake and other areas in the Eastern part of the Kaliningrad region and the Southwestern part of Lithuania;
- a density of population and amount of urban population is bigger in Lithuania's borderland. In Russia's borderland there prevail more population at demographically younger age;
- the different degrees of reclamation of agricultural landscape are appropriate for both borderlands: in Lithuania's part there prevail more reclamated land-uses of agriculture, and Russia's part is more forested;
- the living standards in the Kaliningrad region compared to the indicators of living standards in Lithuania's borderlands. These areas according to the level of socio-economic development and the means of indicators are significantly lower in comparison to average meanings of Lithuania;
- in Lithuania's borderlands prevail more developed services of tourism and tourism infrastructure in comparison to Russia's part. The cultural ethnographical heritage integrated into the network of natural and cultural tourism activities in Lithuania's part as well.
- in Russia's rural border areas there are not any clear specialization of economic activities and governmental initiatives for development of infrastructure, what suppresses the potential activities for spatial interaction on the both sides of the border.

Nonetheless, the increased geopolitical tension in the border regions between Lithuania and the Kaliningrad region, the potential for cross-border cooperation is being seen in many prospects related to promotion of cultural cohesion and cooperation in different economic sectors and their branches. An intensification of spatial interaction depends on conducted researches and gained outcomes. The knowledge on neighboring border regions, it leads towards an efficient regional cohesion and it solves different socioeconomic, cultural, and spatial problems. International tourism and its development is an efficient set of activities that imply to understand an importance of inter-regional cooperation in order to reveal common values and identities in border areas. Elaboration of the international tourism reciprocal activities, it creates the conditions for better spatial interaction of neighboring border regions, as these regions are becoming more integrated in international policies, and they indicate the effectiveness of cross-border cooperation.

Factors of Attractiveness for International Tourism Development in the Curonian Lagoon Region and the Problem of Spatial Interaction

Although, the region of the Curonian Lagoon is homogenous from the point of view of historical processes, however, the natural environment of the region is different on the both coasts, i.e., on the continental part and the Curonian spit. In spite of close geographical proximity of the coasts, there are also different and similar statutes of nature protection. In the Curonian spit there are established two national parks (one in Lithuania's part, and the second one in Russia's part). In the continental part of the lagoon in Lithuania's part established the Nemunas delta regional park, and in Russia's part the nature reserve of the Nemunas delta. The factor of border plays a significant role in the entire coastal region as the boundary between Lithuania and Russia is under strict supervision and control by modern technologies. Even, the natural and cultural landscapes are vivid and both have a significant continuation on the both sides of the boundary, but the created recreational and touristic infrastructure is in different functioning in order to serve the needs of visitors. There is one common historical place and the similar structures of landscapes, but social environment and facilities of physical and recreational infrastructure differ. This is a reason for limitation of spatial interaction on the both sides of the coastal region. The cultural heritage is best protected in the Curonian Spit on Lithuania's side. The situation of tourism promotion differs in both countries as well. There are differences in numbers of tourists, land-use reclamation, and development of infrastructure and different involvement of natural, historical, and cultural heritage into sustainable development of the region. The principles of sustainability and their implementation also differ due to different levels of developed local economies and infrastructure. Recently, a protection of national-cultural heritage has different scenarios of evolution related to application for sustainable national and international tourism development. The processes of evolution based on national strategies that are not adjusted and sustained between two countries. Nevertheless, the key factors of attractiveness for international tourism development is *DNA* of the region like similar natural environment and common historical past, multicultural and tolerant societies, potential for natural, historical and cultural landscape creation, which are based on regional values and identities. A vulnerability of local ecosystems creates the contradictions for sustainable development. The natural heritage is compound of sandy dunes and sandy coast, hills, forests, meadows, reeds, coasts (capes and bays), islands, shallows, rivers and rivulets, lakes, peat bogs, swamps and wetlands, similar flora and fauna on the both sides of the cross-border region. The historical and cultural heritage both are compound of archaeological heritage, urban heritage, land-use heritage, art heritage, scientific heritage, ethnographical heritage, linguistic heritage. The history of the region related to the following persons as the monarchs of Prussia, Simon Dach, Kristijonas Donelaitis, Vilhelmas Storosta Vydunas, Liudvikas Reza, Herman Suedermann, Friedrich Wilhelm Argelander, Martynas Mazvydas, Immanuel Kant, etc.

A cultural process is a mission of the cross-border region. The continuation of the cultural process and constraints of cultural landscape have to be based on sustainability principles, regional values, and identities. It is possible to capture the important factors of cultural identity, spatial, and functional framework of this process through analysis of the cultural environment of the region. The vision of the good continuation of sustainable development, it has to be placed on the foundation of authentic natural and cultural values and the integrity of them into the process of natural and tourism development in order to increase the attractiveness of inbound tourism, which is a part of global international tourism as well. In Lithuania and Russia within the Curonian lagoon region this should be practically synchronized in accordance with legislative documents on local, regional, national, European and worldwide levels. It has been recommended that a holistic approach on heritage protection should be related to on the following legislative documents: European Landscape Convention, European Cultural Convention, and Convention for the Protection of the Architectural Heritage of Europe, The Valletta convention. On the worldwide level—UNESCO, Ramsar Convention, ICOMOS. The trends of shaping the environment should be consistent with the philosophy of sustainability and sustainable development. The following directions for sustainable development as agriculture, international tourism, promotion of economic activities, entrepreneurship, hospitality services, communications are essential for the prosper development of the cross-border region (Table 3).

The suppressing spatial interaction between Lithuanian and Russian parts of the Curonian lagoon region is due to the land and water boundary, which due to implied geopolitical tension becomes a very important political and physical barrier for spatial interaction between both countries. This boundary rather distinguishes than unifies a space that has a common historical past and homogeneous landscape. Therefore, a dialogue for simplified order of border crossing for international tourist, it purposes in this region to elaborate the starting points for ensuring the spatial interaction within cross-border region.

Table 3 Swot analysis

1.1. The Curonian lagoon coastal and cross-border region is rich in history and values, and it has a plenty of objects of natural and cultural heritage	**1.2.** Neither in Lithuania nor in Russia, there is no elaborated strategy for the Curonian lagoon region as for specific natural and cultural territory
1.3. To elaborate the strategy and action plan for natural and cultural heritage protection based on the historical specificity of the Curonian lagoon region	**1.4.** The statewide decision on cultural heritage protection and for protection of separate objects, it depends on political decision, but not on historical, regional values of objects
2.1. There are vivid reflections of Lithuanian (Samogitian), Kurshenieki and German culture in coastal region	**2.2.** A cultural identity of Lithuania Minor is stronger than the topicality of the Curonian lagoon region. There are obvious differences of landscapes and different level of generated infrastructure in comparison to the continental part of coastal region and the Curonian spit
2.3. To promote elaboration of identities by the means of integrating cultural values into joint cultural tourism products and create synchronized tourism infrastructure	**2.4.** The formation of the Curonian lagoon as a separate cultural region has to small value on national scale. Thus, the region not recognized in regional formation and integrated spatial development policy
3.1. Elaborated laws and regulations on protection of cultural and natural heritage are sustained and sufficient in both countries	**3.2.** The principles of sustainability and sustainable protection on cultural heritage are mentioned in the documents, but they are not seen in the practice
3.3. Due to the sensible ecosystems and historical specificity of cultural heritage, the decision making for regional development has to be based on the principles of sustainability	**3.4.** The administrative power of national and regional parks is not efficient to ensure the principles of sustainability on natural heritage protection, except an alternative—a minimal human intervention in natural landscape
4.1. The topicality of cultural and natural heritage protection is involved in national, regional and local development strategies, action plans of municipalities	**4.2.** The topicality of natural and cultural protection in national, regional and local development strategies of municipalities is not unified, adjusted or integrated
4.3. The municipalities has to agree and share the decisions on initiative of cultural heritage protection. Good examples provided by national parks	**4.4.** The strategies and action plans by local authorities perceived and accepted as formalities only
5.1. The regional agents for supervision of cultural heritage objects are the sections of cultural heritage protection either by the Departments of Culture in Lithuania or in Russia	**5.2.** Due to the insufficiency of state financial assignments, there is a lack of qualified specialists, inspectors for controlling and supervision of cultural objects protection in both countries
5.3. The knowledge on cultural heritage protection has to be shared among the specialists of public administration	**5.4.** Due to the lack of state financial assignments for cultural protection, the role of heritage specialist limited in protection duties
6.1. The Seashore regional park, the Nemunas delta regional park, the Curonian spit national park in Lithuania, and the Kurshskaya kosa	**6.2.** The conflicts between local inhabitants, owners of property in the national and regional parks, make reduced efficiency and reduced

(continued)

Table 3 (continued)

national park in Russia, all of them are active agents in natural heritage (landscape) protection	progress of natural and cultural protection. Contradictions between the state (park) owned and private property
6.3. To develop legislative cooperation model between public and private property for cultural heritage objects in order to ensure solutions of conflicts and sustainable application for objects protection	**6.4.** The conflicts between public and private property rights of cultural heritage objects, they deny the possibilities to use the financial grants of the EU and other national fundings
7.1. The municipalities and their sections of economy and investments, as well culture and tourism information, they cooperate in elaboration of projects of investments related to application of cultural objects for public needs	**7.2.** Due to the lack of qualified specialists, which can be able work with projects, the administrations of municipalities are forced to buy the services of external consultancy, these services are expensive and risky in providing of relevant quality
7.3. The administrations of municipalities have to educate the appointed personnel to elaborate the projects' applications, studies of feasibility and preparation of investments' schemes	**7.4.** Due to difficult order of projects' administrations, projects' eligibility, the administrations of municipalities take off responsibilities and devote the projects implementation for enterprises of external consultancy, difficult procedures of public tenders, forces to accept the offers based on the lowest costs/prices, but not necessarily on economic efficient scenario
8.1. The grants for cultural protection projects are given by EU structural fundings (ERDF), and Norwegian and EEA financial mechanisms	**8.2.** The EU and other fundings take a long time in evaluation and approval of heritage protection projects, there are difficulties in elaboration and proofing of the vitality of projects
8.3. The experts of fundings proceed evaluation of applications faster with suggestions for clarification of heritage project for public needs	**8.4.** The calls for applications is at least once a year and big competition, the preparation of investments' project application, it requires prepared technical documentation that is time consuming (up to 2–3 years)
9.1. According to national legislations, the protection of wooden architecture heritage is a priority in the border regions	**9.2.** The local municipal authorities have not sufficient financial resources for wooden heritage protection
9.3. The elaboration of wooden architecture heritage projects for public needs is a priority measure of fundings	**9.4.** The wooden architecture heritage is destroying, wracking due to a lack of financial support from fundings, and unproper the wooden heritage category in state registers
10.1. The application of natural and cultural heritage objects related to servicing of public needs	**10.2.** The application of natural and cultural heritage objects have not been adjusted to neighboring territories; the duplication of actions, it makes lower integration of objects in tourism infrastructure
10.3. The regional coordination of cultural heritage protection projects, it makes better planning of initiatives for heritage protection	**10.4.** The competition between the municipalities and institutions in seeking to get grants for heritage application projects for public needs, this competition separates from regional cooperation and coordination

Strengths (.1.), Weaknesses (.2.), Opportunities (.3.), Threats (.4.)
Author's own research, 2017

Identification of common natural, historical, and cultural values of the Curonian lagoon region is also an essential point for ensuring the further steps on spatial interaction activities, which affect the revealing and involvement of forgotten or neglected places and values, and their integration into the frames for sustainable development of the region. A continuation of regional development action based on heritage protection and its application for sustainable and international tourism development. Elaboration of joint cross-border projects involving the objects of natural and cultural heritage objects and common values for creation of natural and cultural tourism products in cross-border region, these actions imply an importance of international tourism for strengthening the spatial interaction between Lithuania and Russia, and these actions induce a reciprocal understanding on local, regional and nationwide levels.

Conclusions

Looking back at the historical past, there are many things in common comparing the borderlands of both countries. The most important indicators are the objects of cultural and historical heritage, which make a significant reflection for understanding the common values in borderlands. These values insist a reciprocal understanding and they create conditions for cross-border cooperation among societies that experience different scenarios of social and cultural evolution recently.

A case of Lithuanian–Russian border and borderlands, it still was not discussed very much in detail neither by Lithuanian, nor by Russian or so-called Kaliningradian researchers. Therefore, the comparison of borderlands by different aspects leads towards better understanding of existing problems and ongoing processes between two neighboring countries. Some of valuable researches in the field of social-economic geography, demography, geopolitics, regional development geography, cultural geography, ethno geography, all these have been conducted and recognized as pilot researches in order to promote inter-regional cohesion based on cultural and economic cooperation firstly.

An elaboration of international tourism depends on sustainability of inter-regional relations and inter-governmental political approach. An elaboration of touristic products based on common cultural and historical values, and it promotes and reinforces spatial interaction between cross-border regions. The Curonian lagoon cross-border region is vivid in many issues related to tourism development and protection of cultural and natural heritage. The cultural-historical and geographical studies on the Curonian lagoon region creates a plenty of preconditions for planning of spatially joint international touristic routes, joint recreational infrastructure in order to increase the attractiveness of the region for inflow of tourists. Therefore, promoted factor of spatial interaction between the regions, it ensures the continuity of joint actions for cross-border and transboundary cooperation and spatial integration.

References

Baubinas R, Stanaitis S (2001) A change of population in Lithuania's Borderland in 1959–1997, Part 1, The geographic preconditions of population dynamics in Lithuania's Borderland. Ann Geographicae 34:122–138

Baubinas R, Stanaitis S (2002a) A change of population in Lithuania's Borderland in 1959–1997, Part 2, a change of population and its density in the wards of border regions. Ann Geographicae 35:154–168

Baubinas R, Stanaitis S (2002b) A change of population in Lithuania's Borderland in 1959–1997, Part 3, a change of population and its density in the border districts of Lithuania and the Neighbouring Countries. Ann Geographicae 35:169–183

Baubinas R, Kriauciunas E, Taminskas J (2002) The international problems of natural resources exploitation in Lithuania's Borderlands. Ann Geographicae 35:232–263

Bilchak V, Samson I, Fedorov G (2000) The Kaliningrad's pole of integration. Strategy of development of exclave's region. Grenoble-Kaliningrad, pp 363–379

Bolychev O, Gumenyuk I, Kuznetsova T (2015) Local border traffic and the development of the retail trade in the Kaliningrad region and polish borderlands. Baltic Region 4(26):102–112

Buciene A, Spiriajevas E (2005) Human factor analysis in the cross border Curonian Lagoon region. In: Michalski T, Pacuk M (eds) Coastal regions, vol 9. Gdynia, Pelplin, pp 88–108

Burksiene V, Spiriajevas E (2015). Structure of small and medium sized business, and its territorial distribution in Lithuanian part of the Curonian Lagoon region. Regional formation and development studies (1). Klaipeda University, pp 19–35

Burneika D (2003) Lithuania's membership in the EU and impact on business environment in cross-border regions. Authorized scientific work. Vilnius University Library, Vilnius

Druzhinin A (2016) Russia's coastal zone as a social and geographic phenomenon: conceptualization and delimitation. Baltic Reg 8(2):57–67

Euroregions, Governance and Trans-border Cooperation within the EU (2002) In: van der Velde M, van Houtum H (eds) Borders, Regions and People, pp 104–115

Fedorov G, Korneevets V (2015) Socioeconomic typology of Russia's coastal regions. Baltic Reg 4(26):89–101

Introduction: Current Issues and Debates on Borders and Border Regions in European Regional Science (2002). In: van der Velde M, van Houtum H Borders, Regions and People, pp 1–11

Klemeshev AP (ed) (2004) Development concept of the Kaliningrad Region in the conditions of European-Russian interaction in the baltic. In: Region of cooperation, vol 8 (33), Kaliningrad University Library, 92 pp

Kulesza E, Maciejewski A (2003) Osobowe Przejscia Graniczne. Raport. Wspolnota Kulturowa "Borussia", Olsztyn

Liekis A (1997) Dynamics of Lithuania's Boundaries. A Historical review from the ancient times until present 1. Science Lithuania, 542 pp

Paasi A (1996) Territories, boundaries and consciousness: the changing geographies of the Finnish – Russian border. Belhaven Press, London

Palmowski T (2004) The framework of regional development in cross-border areas of North-Eastern Poland and the Kaliningrad Oblast. Coastal Regions, vol 8, Gdynia, Pelplin, 127 pp

Palmowski T, Fedorov G, Korneevets V (2003) Economic, geopolitical and social problems of co-operation between Kaliningrad and Poland. Coastal Regions, vol 6, Gdynia, Pelplin, 141 pp

Pustovgarov VI (2002) Coastal Zone. Economic-Demographic Aspects of Spatial Planning. Publishing House of the Kaliningrad University, 194 pp

Spiriajevas E (2004a) The structure of tourism' sector and its geographic interactions in the Curonian Lagoon Region. Tiltai/Bridges/Bruecken 4:41–52

Spiriajevas E (2004b) Tourism as perspective direction for cross-border cooperation in the Curonian Lagoon Region. Cross-Border Cooperation, Russian Baltic, pp 186–201

Stanaitis S (1996) The changes of demographic peculiarities in the Curonian Lagoon Region. Ann Geographicae 29:316–328
Szlachta J (2004) The European Union's Cohesion Policy 2007–2013. Opportunities for the regions of the south Baltic development zone, Warszawa
Timothy DJ (2001) Tourism and political boundaries. Advances in tourism. Routledge, London, pp 1–7
Vaitekunas S et al (2001) Social and geographical strengths and weaknesses of tourism development in the Curonian lagoon region. Tiltai/Bridges/Bruecken 1:51–56

Movement to Defend the Białowieża—The Problem of the Białowieża Forest Protection as an Example of a Values Conflict

Małgorzata Bieńkowska, Łukasz Faszcza and Łukasz Wołyniec

Abstract The Białowieża Forest is a part of UNESCO heritage. In 2015, when the Polish Minister of Environment allowed the mass logging in a protection zone of the forest and in the forest itself, as the ecologists proved, he explained these actions as "protecting" the forest from the bark beetle. The plan executed by the Ministry resulted in a deep conflict between the government and the citizens. The article that we present demonstrates a historical overview of the forest management, as well as contemporary conflict relating to this area occurring in Poland. Not only did this conflict appear to be serious for Polish citizens, but it also engaged the EU institutions.

Keywords Conflict · Environmental protection · Białowieża forest

Introduction

Białowieża Forest is a unique forest area—the last remaining lowland forest in Europe. In 1921 the part of the forest was declared a national park—a National Reserve. European bison has become a symbol of the Park—it is the biggest land mammal of Europe. Białowieża Forest became the last refuge to the lowland bison. It was also where the process of bison restitution was initiated, returning bison to life in natural conditions. At present, it is Białowieża Forest where the population of freely roaming bison is the biggest in the world. In its part in Poland there are about 500 bison. Białowieża National Park is the only natural site in Poland being

M. Bieńkowska (✉) · Ł. Faszcza · Ł. Wołyniec
Institute of Sociology and Cognitive Science, University of Białystok, Białystok, Poland
e-mail: m.bienkowka@uwb.edu.pl

Ł. Faszcza
e-mail: lukaszfaszcza@wp.pl

Ł. Wołyniec
e-mail: l.wolyniec@uwb.edu.pl

J. S. Methi et al. (eds.), *Borderology: Cross-disciplinary Insights from the Border Zone*, Springer Geography,
https://doi.org/10.1007/978-3-319-99392-8_3

included in the UNESCO World Heritage Site list. It is also the core-central zone of Białowieża Forest Biosphere Reserve.

Historical Note: Economic Exploitation of Natural Resources in Białowieża Forest Since 1939

Natural landscape of Białowieża Forest has been formed and changed by human activities for several centuries. They were connected with colonization that can be traced back as early as 1,000 years BC (Samojlik 2007, pp. 49–50). Gradual populating process of Białowieża Forest that started in the thirteenth century appeared to be chaotic and uncontrolled. It was hunting and gathering that were the major activities within the population although breeding animals and agriculture were also noted (Samojlik 2007, pp. 43, 55–59). As the areas neighboring with Białowieża Forest were exposed to numerous military conflicts and battles, they were not a favorable place for settlement. In the thirteenth century the region was the property of Volyn dukes. However, in the first half of the fourteenth century it was ruled by the Duke Gedymin, as a result of the conquest (Hedemann 1939, p. 21). Conflicts tearing the elites of the Grand Duchy of Lithuania, frequently turning to civil wars, prevented any settlement actions until the end of the fourteenth century. Besides, Lithuanian rulers were constantly involved in the disputes with Poland. As a result of one of these disputes, in about 1348, Białowieża forests were included in the Kingdom of Poland for a short period. The Polish-Lithuanian Union, set in 1385 in Krevo, made Jogaila (later known as Vladyslav Jagiełło) king of Poland and his cousin Vytautas the ruler of Lithuania- this allowed initiating another stage of Białowieża Forest colonizing. However, this time it was the state that took steps towards the colonization (Tęgowski 1999, pp. 10–16; Bajko 2004, p. 9).

In the fifteenth century, the Kingdom of Poland and the Grand Duchy of Lithuania were united by the common ruler. Within the gradually emerging vast state in Central Eastern Europe there were endless forests. For centuries Białowieża Forest was not economically exploited. It was treated as a unique place where the rulers and monarchs hunted. As a result, the forest stand remained practically unchanged until the end of the sixteenth. It was only at the beginning of the seventeenth century when the decisions of more serious exploitation of Białowieża Forest were made (Samojlik 2007, s. 86; Bajko 2012, s. 11–12). Those interested in wood procurement were obliged to sign a contract with state administration. This agreement allowed the monarch to lease a certain area of the forest to the contractor who procured the appropriate amount of timber at his own expense. However, forest stand in Białowieża Forest was not affected due to the intensified exploitation (Hedemann 1939, pp. 251–253). In the second half of the eighteenth century, an essential reform related to forest stands in the properties belonging to state treasury was introduced, encouraged by Antoni Tyzenhaus, the treasurer of the last king of Poland-Stanisław August Poniatowski. According to the regulations introduced in

1765, person signing the leasing agreement to use forest areas, was not entitled to the right of their free exploitation. Abolishing this right was a good solution, as the logging done by the contractor in the uncontrolled manner could result in the excessive abuse. It was Antoni Tyzenhaus who decided that it was exclusively state administration and its representatives that could sell the timber to those being interested (Hedemann 1939, pp. 254–257; Szafer 1917, p. 12).

Yet, this was not the only action of the treasurer. Natural resources of the Białowieża Forest had been desirable among the local population for many years. The right to enter the forest served as a kind of reward or privilege granted for the dedicated service to the monarch or the country, confirmed by an official act in a written form. It frequently happened that some people not being in possession of the act entered the forest following the legend of the act having once been granted to them. However, it was inconsistent with the existing law. In the second half of the sixteenth century, the verification of the privileges allowing using the natural resources of Białowieża forests was organized, during the reign of Sigismund Augustus. Antoni Tyzenhaus made this issue to be examined again, not merely analyzing the particular cases but revoking the privileges (Hedemann 1939, pp. 102, 132–134).

Exploiting the Białowieża Forest was not limited to procuring timber. For years, Białowieża forests have been the source of charcoal and the potash (kind of ash). These products were obtained both in large-scale (according to lease contracts) and in small amounts. Białowieża Forest was also the place of such supplies as iron-ore, tar, huge amounts of honey; it was also the place for cattle grazing, fishing, and hunting. Hunting was yet partly limited and subjected to numerous regulations making it legal to very few people (Hedemann 1939, pp. 273–293).

In 1795, after the Partitions of Poland, this country disappeared from the maps of Europe for 123 years. The area of Białowieża Forest became a part of Russia and Prussia. Vast part of Białowieża forests became the property of tzars for over a hundred years. Russian generals fighting against Poland demanded gratifications from the Tzarina Catherine II; they were usually rewarded with the land property, being previously the property of Polish Treasure. As a consequence, a part of Białowieża Forest found its new landlords who did everything in order to obtain great profits from their new properties; it meant intensive exploitation of its timber supplies. Fortunately, soon the whole area of Białowieża Forest became the property of Russian authorities again (Więcko 1963, p. 299, Samojlik and Jedrzejewska 2005, pp. 64–66). In 1811, the forest was partly devastated by the huge fire; several thousand people were sent to fight against the disaster yet they did not manage to win. These were only October rains that finally stopped the fire. The scale of devastation was so serious that it took 8 years to re-establish the site destroyed by the fire. In 1821, Alexander III declared this area not to be subject of any economic “colonization”; this situation lasted until 1838, when the Imperial Navy demanded the source of timber to construct ships. This resulted in intensive logging, begun in 1839. However, later it was revealed that most of the timber turned out to be useless for ship construction, not mentioning an incredibly high cost of transport. The whole procedure was chaotic and it well reflected thoughtless activities of tsar's administration. In the 1840s, new regulations of forest management were introduced, defining average annual exploitation of the forest. These regulations followed more rational principles.

Although the logging, lasting until the end of the nineteenth century, was based on economic principles, it did not result in significant deforestation of Białowieża area. In 1897, Alexander III ordered to stop forest clearing in Białowieża area (Więcko 1963, pp. 324–338).

During the World War I, Białowieża area suffered heavy losses. In summer 1915, German army, having defeated the Russian soldiers, seized the area of Białowieża forests. Local towns and villages were destroyed as a consequence of military actions but the forest itself did not suffer major losses (Tomkiel and Markowski 2016, pp. 21–45).

Long-lasting military conflict meant an increasing demand for raw materials. Apart from coal, metal, and food, it was timber that was needed in large quantities; it was necessary to construct the revetments or rail ties. Białowieża Forest, being according to German soldiers, the endless source of raw materials, soon became subject to intensive exploitation. It took only 3 years to export the amount of timber equal to 1% whole forest stand. It was possible due to huge investments done by German authorities, including narrow-gauge forest railway and lumber mills. Georg Escherich, a famous German explorer, supervised the whole process (Rivoli 1918, pp. 73–86; Nikitiuk 2003, pp. 21–32).

The end of the World War I did not mean the end of exploitation of Białowieża Forest. The defeat suffered by the German army and the newly emerged independent Polish state resulted in the increasing demand for timber necessary to rebuild towns and villages destroyed during the war. Again, it was Białowieża Forest being the source of raw materials. In 1924, Polish government searching for additional sources of revenue, leased a vast part of the forest to British trade company "Century" for 10 years. This was a disastrous period for Białowieża Forest, with predatory exploitation, devastating the forest more than it was done by the German army during the World War I. In 1929, Polish government, managed to break the contract, due to the fact "Century" violated its terms and conditions; the government was also significantly pressed by the public opinion (Więcko 1984, pp. 34–45).

The Problem of the Białowieża Forest Protection as an Example of a Values Conflict

The question of Białowieża Forest protection is under public debates not only in Poland but also in European Union and other international organizations. A few years ago emotions were inflamed by central government's proposals to expand the area of the Białowieża National Park, or by the search for another ways to strengthen protection regimes in one of Europe's last complexes possessing the primary forest characteristics.

The debate around the Białowieża Forest can be reduced to the dilemma of what kind of protection would be optimal to preserve the natural values of this forest complex.

On the one hand, there is a suggestion for the *active protection* of the forest, which involves intervening in natural processes. These could be done for example by pest[1] control, felling or planting, which involves some form of forest management.

In this perspective the forest is treated not only as a unique habitat for rare species of plants and animals, but also is seen as a reservoir of resources that can be obtained for business purposes. We may include to the supporters of this concept a large part of foresters, local authorities, as well as some groups of residents. A similar point of view appears to be represented by current leadership of the Ministry of Environment.

On the opposite side there are the supporters of the *passive protection*, which comes to total refrain from conducting not only forest management, but also taking any treatments or sanitation. This is based on the conviction that nature "could take care of itself". This means that if we observe any phenomena that may be treated as a disorder from the point of view of forest management's effectiveness, in this perspective they are treated as the part of natural processes, in which we should not interfere. We can even hear the opinion that Białowieża Forest should stay "wild" and therefore should be "untouched" by human hand.

In the face of local authorities' opposition to government's proposal to expand the national park to the whole area of the Polish part of the Białowieża Forest, different ways of forest protection were proposed, including radical reduction of harvesting (The Forest Management Plan adopted in 2012 assumed the reduction of felling form over 100 thousand cubic meters per year to 48.5 thousand cubic meters annually) and also the ban on felling of older than a hundred years forest stands, which significantly slowed down timber harvesting. Since then the Polish part of Białowieża Forest had the strictest protection regime in its history.[2]

[1]It should be noted that the term "pest" is relative and depending on the parties to the dispute. For example, the bark beetle is treated as a "pest" primarily by the foresters, since this species, attacking spruces, contributes to the death of trees and makes tangible (and quantifiable) economic damage. Supporters of passive protection (which means the refrain from intervention in natural processes) generally do not treat bark beetles in these categories. For them it is another species living in the ecosystem and its gradations are treated as "normal" natural process. A similar problem with the "dead wood" also requires an explanation. For the large part of the *active protection* supporters it is a manifestation of wastage of forest resources. On the other hand, the organizations opting for *passive protection* of the forest see it as a habitat for many valuable species (e.g., insects, fungi, plants, etc.).

[2]Note that Białowieża Forest was divided by the state border after the war in 1945.

However, the FMP[3] did not contain a precise distribution of limits of harvest between three forestry managements operating there, which was later noted by the Supreme Chamber of Control in its report.

There is no doubt that many groups were not satisfied with the restrictions laid down by the Ministry of Environment. We point not only the foresters or local governments, but also, paradoxically, some environmental organizations—the first saw further restrictions and impediments—the latter, still too weak regimes of protection. So we can say that, although no one had fully accepted these solutions, the parties functioned as if "in abeyance" and the conflict appeared to be severely limited.

On the one hand, areas outside the national park have received the best protection status in the history—when the expansion of the BNP became impossible, the government (Ministry of Environment) tried do find different solutions. On the other hand, a part of the opponents of tightening protection regimes (e.g., expansion of the national park, reduction of the felling, etc.) in the face of gradual reduction of forest management and the introduction of the so-called "*non-national-park forms of protection*", although dissatisfied, seemed to be reconciled to some extent with that fact. The conflict therefore has not been resolved, but only reduced its intensity, however "strategic initiative" was on the side of supporters of *passive protection*, and the groups opting for forest management (*active protection*) were pushed onto the defensive.

At the same time frustration has grown in local communities because they were concerned that the decision of tightening the environment protection was taken with insufficient attention to local communities' needs.

Everyone Loves the Forest

Frequently we could meet the opinions of the residents that the cost of nature conservation was passed on local communities. They often drew attention to the limitations related to the deterioration of protective regimes, which would result in difficulties in investments in the area.

In addition, during interviews, respondents repeatedly emphasized their feeling of "communication exclusion", arguing that the government listened only to the voice of environmental organizations and scientists, completely overlooking arguments of residents and foresters. Representatives of the latter group often feel affected by the opinions of environmentalists, according to which foresters allegedly are "not protecting the forest". In such situations, respondents usually cited a number of examples that refuted this thesis, and noted that for many years the foresters made a great contribution to the protection of the Białowieża Forest.

[3]Forest Management Plan.

On the other hand, sometimes we meet with naming activists "*eco-terrorists*", who delight in spectacular actions and "prefer the worms than the interests of the residents".

Interestingly, the foresters, residents, scientists, and ecological activists declare that they "love the Forest" and wish it a better future. The problem is that each of these groups differently understand the word "good". Some of them would like to take immediate action to stop the bark beetle gradation (by cutting out infected spruces), the others would prefer the nature to govern itself without any human intervention.

The examples mentioned above seem to confirm the thesis of organic lack of mutual understanding between the main parties. Practically all actors involved in the case drew the attention on this issue. There is the question, whether this misunderstanding stems from the lack of willingness to listen to the arguments of adversaries, or maybe it is primarily due to the use of incompatible languages by the both sides of conflict. On the one hand we have an utilitarian attitude towards nature, which largely focuses on the possibility of discounting these unique resources. On the other hand, the nature is treated more as autotelic value, which should defend itself and does not require any human intervention. Is it therefore possible to achieve a compromise, which would be stable and satisfying all parties?

Political Change and the Forest Białowieża at Present—Political Conflict

In autumn 2015, a majority government was formed in Poland as a result of the parliamentary election that was won by Law and Justice, a conservative right-wing political party that was able to form majority government. Election promises that were declared by this party were based on the social project "Family 500+", providing families with two or more children with financial support of PLN 500. Soon after the election, the project was implemented, being still in operation. Although its social advantages are unquestionable, it is still the subject to numerous economic controversies, related mainly to its possible financial sources that need to be defined.

Jan Szyszko, the Minister of Environment in the government, has found the provisional solution to economic problems. He is in favor of the opinion declaring Poland to own large timber resources in Białowieża Forest, worth "millions". When he was harshly criticized by ecological organizations he went on to insist on Białowieża Forest destruction caused by spruce bark beetle and, consequently an urgent need for forest clearance. There was a strong objection to the ideas expressed by the minister, both in ecological circles and among the Internet users. Social media has helped to organize numerous protest actions since March 2016, such as the March of the Ents, trying to make the minister and the government realize the fact of the forest value being something more than the subject of economic calculations.

After the parliamentary elections in 2015 and the change of the ruling party, the Minister of the Environment began to withdraw from the solutions adopted by his predecessor. This resulted in a significant increase in the intensity of the conflict, but this time the ecological organizations have been pushed onto the defensive. However, in the face of the changes forced by the ministry, these organizations have been mobilizing to halt negative, in their opinion, changes.

At the moment, it seems to prevail a foresters concept of *active protection* of the forest, consisting mainly of removing bark beetle infested stands, which was criticized by environmentalists and journalists. Sometimes the object of the protest is not the mere fact of the felling, but the way it is conducted, e.g., using the most invasive methods—special tree harvester machines (Ludwiczak 2016).

The Collision of Materialist and Post-materialist Values

That conflict can be analyzed as an example of the collision of materialist and post-materialist values resulting from R. Inglehardt's concept. At the core of this theory lays the concern that the orientation towards the values is linked to the level of satisfaction of material needs—this is reflected both in individual attitudes and also in differences between societies.

According to this concept, higher needs and values are revealed after satisfying the primary needs. They may be for example the values concerning democracy, freedom, equality, but also ecology. This may result in more autotelic valuation of the natural environment, opposing to more materialistically oriented groups, for which priority is to ensure social security and their attitude to nature is more utilitarian. This hypothesis may explain the difference between the attitude to Białowieża Forest of local communities (which may represent more materialist orientations) and ecological organizations (with stronger post-materialist orientation). This hypothesis should be verified by further research.

Conclusion

The main part of this article was written during the intense phase of the Białowieża Forest conflict. It began when the foresters intensified "active forest management" to fight the bark beetle gradation, which had been present in this area for several years. Until 2015 active forest management was conducted to a very limited extent. This approach changed after the parliamentary elections when Jan Szyszko became the Minister of the Environment.

The introduction of heavy equipment and the drastic increase in timber harvesting led to the escalation of the conflict and the activation of ecological and scientific environments. Eventually, the European Union institutions were involved in the dispute. As the consequence of internal proceedings The European

Commission engaged the Court of Justice of the EU. The statements expressed by Polish Minister of the Environment meant he was fully convinced that the activities he performed were correct.

Nevertheless, following the decisions taken by the ruling party and the reorganization of the government at the end of 2017/beginning of 2018, Jan Szyszko was replaced by Henryk Kowlaczyk, who declared his full respect for the rulings of the CJEU.[4] At the same time, the felling in the Białowieża Forest was stopped. In April 2018, the Court of Justice stated that Poland violated European law, especially The Habitats Directive and The Birds Directive, and thus the felling was illegal.[5]

It seems that this decision should finally close the issue of the legality of forestry activities, but now a new conflict seems to open. The foresters in accordance with practice of forestry art carry out planting the new trees on the clearcuttings. Ecological environments postulate the natural regeneration of the forest. They refer to the processes that took place after a drastic felling carried out in the years 1924–1929 by the British company "The Century European Timber Corporation". Most of the cuttings underwent spontaneous renewal and the natural processes take place in the forest again.

Scientists associated in the "Science for Nature" initiative cite a number of arguments that planting poses more problems than benefits.[6] Therefore, it seems that although there are bases to believe that the felling in the Białowieża Forest has been effectively stopped, the conflict itself has not yet been resolved and there are no final solutions for protection of the Białowieża Forest as well as no basis for local communities development was proposed.

References

Bajko P (2004) Kalendarium białowieskiego żubra. Białowieża

Bajko P (2012) Echa starej Białowieży. Białowieża

Hedemann O (1939) Dzieje Puszczy Białowieskiej w Polsce przedrozbiorowej. Warszawa

http://curia.europa.eu/juris/liste.jsf?language=en&td=ALL&num=C-441/17

Ludwiczak U (2016) Ciężki sprzęt w Puszczy Białowieskiej. Ekolodzy: wycina, ale i niszczy, Gazeta Współczesna (http://www.wspolczesna.pl/wiadomosci/hajnowka/a/ciezki-sprzet-w-puszczy-bialowieskiej-ekolodzy-wycina-ale-i-niszczy,10605046/)–access 31.05.2018

Nikitiuk B (2003) Z dziejów Hajnówki i jej okolic (1915–1939). Hajnówka

Rivoli J (1918) Puszcza Białowieska w administracji wojskowej niemieckiej, Przegląd Leśniczy nr 3

Samojlik T (2007) Antropologiczne przemiany środowiska Puszczy Białowieskiej do końca XVIII w. Kraków

Samojlik T, Jędrzejewska B (2005) Koniec epoki, (in:) Ochrona i łowy. Puszcza Białowieska w czasach królewskich, (ed.) Samojlik T. Białowieża

Szafer W (1917) Opis Puszczy Białowieskiej wedle dzieła E. Eichwalda, Sylwan, nr 1–6

Tęgowski J (1999) Pierwsze pokolenia Giedyminowiczów. Poznań

---

[4]CJEU—Court of Justice of the European Union

[5]http://curia.europa.eu/juris/liste.jsf?language=en&td=ALL&num=C-441/17.

[6]Żmihorski et al. (2018)

Tomkiel M, Markowski M (2016) Bielsk 1915. Zapomniana bitwa. Białystok
Więcko E (1963) Dzieje Puszczy Białowieskiej do roku 1918. Kwartalnik Historii Kultury Materialnej, R.XI, nr 2
Więcko E (1984) *Puszcza Białowieska*. Warszawa
Żmihorski M, Chylarecki P, Orczewska A, Wesołowski T (2018) Fakt: Nasadzenia szkodzą Puszczy Białowieskiej, https://naukadlaprzyrody.pl/2018/04/03/fakty-nasadzenia-szkodza-puszczy-bialowieski/–access 31.05.2018

Potential, Problems, and Challenges of Joint International Master Programmes: Case-Study of the Joint Norwegian-Russian Master Degree Programme in Borderology

Inna Ryzhkova and Jan Selmer Methi

Abstract The article is devoted to one of the most important forms of internationalization of higher school—student and teacher academic mobility reflected within joint master programmes. The authors analyse in detail the specificity of developing and implementing the Norwegian-Russian Master Degree Programme in Borderology, the participants of which are the University of Nordland and Murmansk State Humanities University. Cross-border cooperation between Universities, developing in the dialogue of different cultures, is interpreted as "the zone of innovations", within the framework of which a new multicultural educational environment and new teaching methodology appear.

Keywords Internationalization of higher school · The bologna process
Joint degree programme · Transnational education · Dialogue of cultures
Academic mobility

Introduction

Reforming of system of higher education in the view of the requirements of Bologna Process, globalization and internationalization, defining vectors of development of educational systems of the leading countries of the world, is

---

This article was first published in *Sovijus—Interdisciplinary studies of culture*. T. 3, Nr. 2, 2015, Vilnius, Lithuania.

I. Ryzhkova (✉)
Murmansk Arctic State University, Murmansk, Russia
e-mail: innaryzhkova@yandex.ru

J. S. Methi
Nord uiniversitet, Bodø, Norway
e-mail: jan.methi@nord.no

J. S. Methi et al. (eds.), *Borderology: Cross-disciplinary Insights from the Border Zone*, Springer Geography,
https://doi.org/10.1007/978-3-319-99392-8_4

demanded a special close attention to the experience of implementation of the international joint study programmes. Joint programmes, both Master Degree Programmes and Bachelor Degree programmes, are crossing borders and have to be carried out in the field of dialogue of various cultures.

In the light of a basic novelty and innovation of the form of such cooperation of the Universities itself,[1] and also in the light of absence of the uniform universal scheme of the organization of the international joint study programmes, it is necessary both the detailed analysis of the potential inherent in themselves at the level of national and regional contexts and also identification of institutional mechanisms of their implementation and a complex of the risk factors which as a whole one way or another are reflected in a quality of training and development of the international cooperation.

The aim of the authors of this article is to show specifics of the joint Norwegian-Russian Master Degree Programmes in Borderology, which is deeply innovative according to both the content and the form, reflecting qualitatively new stage in development of the international cooperation of higher school of Norway and Russia. According to the fair remark of N. Burkel, S. Tvorgova and S. Shenderova, "today the international joint study programmes are interesting experimental platforms keeping innovative forms of the international cooperation, teaching and studying" (Burkel et al. 2014: 11). Such experimental platform for the joint Master Degree Programme is being analysed by the authors of this article on the philosophical plan together with a focus on comprehension of border as a phenomenon in its broadest understanding.

According to the report "Innovations and changes in transnational education. Joint study programmes between the European and Russian institutions of higher education", carried out for Representation of the European Union in Russia, it has showed that the most part of the European-Russian joint study programmes centre on such areas of knowledge as social sciences, business and law, where management and economy are the most popular within this category.[2] The second place is taken by the joint study programmes in the field of the engineering sciences and also manufacturing and construction sectors (23%); next the humanities and art (10%) and etcetera.

The Russian institutions of higher education cooperate most often with the partners from Germany and France (nearly a half of all study programmes accounts for the total), then it is the partnership with the institutions of higher education in Finland and Great Britain, which is 20% of the total of all joint study programmes. Finland of all EU countries adjoining to Russia advances most actively cooperation in the sphere of the higher education. It should be noted that it is small and only 1% of the total of joint Russian-Norwegian study programmes. English is the most

[1]The first joint European-Russian study programmes appeared in the early nineties, and now this form of interuniversity cooperation develops actively at the qualitative new level.

[2]In the mentioned report the data of 317 joint programmes between the Russian and European institutions of higher education was collected: 133 were already revealed in the research of 2010; 184 new programmes were found in addition.

widespread language of studying the joint study programmes. The most of the joint study programmes are implemented in the large centres: Moscow (44%), St. Petersburg (16%) and institutions of higher education in Siberia (13%). Northwestern Federal District incorporates only 6%. The academic mobility of students within joint Master Degree programmes emphasizes once again asymmetry of its ingoing and outgoing forms: generally the Russian students go to Europe for studying. But there is still rather difficult to attract the European students for studying in the institutions of higher education in Russia (Burkel et al. 2014: 13–14, 69–70, 78–80).

The Definitions of the Concepts "Transnational Education" and "Joint Study Programmes"

To be shy of terminological welter it is necessary to give the definitions of concepts used in the article. Thus, Council of Europe/UNESCO considers transnational education as «all types of higher education study programmes, or sets of courses of study, or educational services (including those of distance education) in which the learners are located in a country different from the one where the awarding institution is based. The institution or programme in question may belong to the national education system of another country, or it may be independent of any national system». European University Association giving priority to the quality of education suggests the following understanding of joint study programmes: «those study programmes which are developed and implemented jointly by several institutions in different countries» (EUA 2006: 9). In JOIMAN 2009: Glossary the following definition is given: «A joint study programme refers to situations when two or more institutions cooperate on a joint study programme that leads to a degree at one of the partner institutions. Two or more institutions collaborate on a study programme leading up to a degree at one of the institutions. One institution is responsible for the degree and issues the degree diploma. This institution has academic responsibility for the contents of the study programme. The course descriptions must be approved by the institution that issues the degree, and this institution is responsible for assessing the basis for the diploma in relation to the descriptions of learning outcome in the programme description. The institution issuing the diploma is responsible for academic approval of the course and for the quality of the programme» (JOIMAN 2009). In 2011 European Consortium for Accreditation in Higher Education confirmed that there is no official definition of joint study programmes, but according to the Commission some elements should be in, for example:

- Programmes should be developed and approved jointly by several higher education institutions;
- Student stays are of comparable length;

- Periods of study and exams passed at partner institutions are automatically recognized;
- Professors of each institution teach at the other institutions and form joint commissions for admission and examination;
- Graduates obtain either the national degrees of each institution or a degree jointly awarded by them (ECA 2011: 7).

The History of Cooperation of the Universities Within Master Degree Programme "Borderology" (Joint Master Degree Programme in Borderology)

The first ideas concerning the programme "Borderology" and potential possibility of the development and implementation of the joint Master Degree programme were presented by one of the authors of this article Jan Selmer Methi at the meetings with Russian colleagues in Kirkenes in 2010. The impressions of these first meetings were very precisely expressed in one of the interviews by the professor, Doctor of Philosophy, rector of Murmansk State Humanities University A. Sergeev: «It took us a long time to grasp the specific character of the approach of our Norwegian colleagues to this course, in fact, things only began to clear up when we invited a group of Norwegian students to Murmansk and had a chance to meet them face to face and talk to them. The matter is that the programme was based on a specific approach to teaching whereby students are asked, at the very beginning of the course, to write an essay about a problem they find particularly urgent or an event that has deeply impressed them. Proceeding from such a problem, an existential one, in fact, students taking part in the Master's Degree programme become gradually acquainted with the world of theoretical knowledge, something that severs as a basis for reflective experience enabling them to broaden their initial understanding of the problem and of their own subjectivity, leading them on and on to their crowning achievement that must presumably take the form of a Master's thesis. Nothing could have been more different from the standard practice of teaching philosophy at Russian universities to which we were all accustomed and which is based on the exact opposite of this procedure: the Russian student is 'plunged' without the slightest delay into the world of theoretical reflection and the grand philosophical systems. I remember very well even today how difficult it was at first to realize that you were dealing with the existential roots of thinking, and once that was clear, how difficult it was to put your trust in them, to accept the subjectivity of every student and then make an earnest effort to understand it. I remember very well how it all set me reflecting on the roots of philosophical thinking itself, as it was practiced in ancient Greece, and how this situation was now repeating itself before my own eyes! This was an unforgettable experience, and it was worth all the difficulties that we had to face at different stages of the project».[3]

Revisit the beginnings of cooperation of the Norwegian and Russian institutions of higher education, the feeling of unlikeliness of real implementation of the joint Master Degree programme based on the philosophical issues within so unusual methodology to the Russian pedagogy of higher school is remembered. It should be noted that the previous significant amount of the international cooperation of MSHU (the regional institution of higher education located in a border-zone, which considers its mission as promotion of the interests of the Russian Federation in Barents Euro-Arctic region) did not include within itself such procedures, based on especially philosophical matrixes. But even if the logic of implementation of other joint Master Degree programmes included elements of the innovations, it would not be so cardinally expressed, forced to change the general dialoguing in the course of educational process (Ryzhkova 2013). It was necessary to combine both the Russian and Norwegian traditions of educational system, taking into account experience of transformations of the educational fields connected with the implementation of the principles of the Bologna declaration. Besides, many questions were also generated by specifics of recruiting of the Norwegian and Russian students for studying such joint study programme, because the recruiting has to be carried out within different cultures and has to consider not only the specifics of motivations, but also experience of the previous studying of the Norwegians and the Russians. Besides, the existence of an Bachelor's Degree or Specialist Degree diploma, the advanced level of proficiency of English, it is required from the students the existence of the seniority at least 3 years, allowing to speak about a gain of one or another practical knowledge experience, that is the necessary for participation in the programme.

It should be noted that initially the participation in the programme does not suppose that the graduation with one more diploma, in this case—Master's Degree, will surely cause of change of employment or direct promotion that what the Russian undergraduates focus on traditionally. During interviewing the future undergraduates students, the professors carefully were explaining a purpose of the programme: mainstreaming of the accumulated experience of professional or private life, comprehension and reflection of it on the basis of the formed ability "to get used" to the text and to put those questions to himself or herself which become the basis of personal development of the students in the future.

[3]This interview was recorded by Inna Ryzhkova in February, 2015.

Supporting Reasons of the Joint Norwegian-Russian Master Degree Programme in Borderology

The numerous subsequent meetings between Norwegian and Russian philosophers, our joint seminars and discussions have made it clear that the phenomenon of the border should be understood as broadly as possible, and that it is also how it should be accepted. To be precise, it should be seen not only as something that separates and distinguishes one thing from another, but also as a real place and intellectual topos for conjoining what is different, one where the experience of the Other first becomes a relevant and important, and then even a necessary moment of your own, personal experience. In this sense, the border serves as a mirror, or maybe even a prism, with the help of which man and society can view themselves through the experience of the Other; it gives you a kind of experience that reflects your self-understanding and your understanding of the phenomena of life, culture, language, and society. The experience of exploring the border enables us to understand the universal and the particular in our own life and in that of our neighbours across the border, whereas a more profound reflection on this phenomenon helps us to discover new dimensions of Being and of events on the border.

It is easy to see that everything discussed above may provide a number of research topics for any University or academic institution, no matter whether it is based in Russia or Norway, in the North or in the South. However, the very experience of co-existence on the opposite sides of a border not only gives an additional impulse for discussing such topics, and a stable one at that, but also provides a 'marker' for events occurring on the border and serves as a catalyst for the understanding of Being on the border that can now be regarded as more than just a mental 'product'. It is impossible to deny the importance of the interest that underlies theoretical reflection, but no less important is the institutionalization of this interest which enables it to be incorporated into a paradigm of reflection and to gain strategic stability thanks to being included in a schedule of joint, international meetings and classes.

Joint Master Degree Programme in Borderology as an Experience of Institutionalization of Interests

The first experience of such institutionalization has taken the form of the Norwegian "Master's Degree Programme in Practical Knowledge (Borderology)"[4] jointly taught in English by Norwegian and Russian academics, mostly philosophers, many of them from the University of Nordland (Bodø) and Murmansk State

[4]Joint Master Degree in Borderology uses the model for Master in Practical Knowledge which was developed by University of Nordland in 2000.

Humanities University. This Norwegian-Russian Master's Degree Programme (2013–2015), realized in the Russian-Norwegian border zone, with such locations as the cities of Nikel and Murmansk (Russia) and Kirkenes (Norway), has resulted in a series of meetings and the establishment of multilateral contacts. International research seminars and conferences have been held at Murmansk State Humanities University (2013–2015) and at Kant and Bakhtin Institute in Nikel (2014). The establishment of this latter institution can be seen as another important step in the institutionalization of the interest in the study of the phenomenon of the border and that of the cross-border cooperation between the two peoples. The names of I. Kant and M. Bakhtin in this case are not to be taken as indicating certain methodological commitments; they rather serve as symbolic constants defining a spectrum of problems, widely differing in terms of their genesis that can be discussed together in the spirit of dialogue and peace. In this respect, the figures of Kant and Bakhtin acquire special significance and function as a kind of textual matrices for dialogical rather than monological constructions.

It should be added that thanks to the seminars and conferences the combined philosophical potential of the University of Nordland and Murmansk State Humanities University was significantly augmented by our colleagues from the Arctic University of Norway (Tromsø), Saint Petersburg State University, Northern (Arctic) Federal University (Arkhangelsk), the University of Helsinki and the University of Lapland (Rovaniemi, Finland).

Master Degree programme in Borderology, being implemented within the consortium of two partners, is carried out in the part-time form of study: as a rule 6–7 three-day assemblies during one academic year as on the territory of Norway and Russia. Study for the Russian students is on a fee-paying basis: they pay only annual fractional semester registration fees and independently cover all expenses on the territory of Russia, at the same time all their living expenses on the territory of Norway are covered by the Norwegian party. It should be noted that initially the Ministry of Foreign Affairs of Norway gave the grant for supporting of this programme with 4 million NOK (Norwegian kroner).

According to the European legislation, the Master's Degree Programme includes 120 ECTS. Term of study of the Joint Master Degree Programme is 4 years (provided that it is part-time form of study) which are divided as follows: the first two years students visit the lectures, at the same time they write the essay, representing results of their reflections at the seminars; the aim of the third and the fourth year of study is a writing of the Master thesis, a larger philosophical thesis on 80–100 pp based on a self- chosen task within the theme borderology. The philosophical thesis will receive an independent grade and the result of the thesis counts as 40% of the combined final grade.

There are no examinations as in the standard Russian tradition in this programme: at the end of the first and second study years the students write the so-called final essay on the basis of which the final assessment is given (from A to F—according to the Norwegian legislation): part-examination 1: "Writing subjectivity" (30 ECTS) and part-examination 2: "Borderology" (30 ECTS). "Writing subjectivity" means a philosophical essay from the area of practice, an individual home-based exam. Results count as 25% of the combined final assessment. "Borderology exam" means a philosophical text connected to own experience and one of the modules from first and second semester, an individual, home-based paper. Results also count as 25% of the combined final assessment. Students who are absent for more than 20% of the total semester periods would need approval from the academic personnel before he or she would be allowed to register for examinations.

Taking into account that the programme is implemented within the "model of joint development and implementation", functions of full Universities—partners are distributed. The Faculty of Professional Studies of University of Nurland is responsible for the implementation of the first study year, beginning with the implementation of educational modules and finishing with a final exam. Murmansk State Humanities University is responsible for the implementation of the second study year that is written in the bilateral partnership agreement, signed by the rectors of both institutions of higher education.

In according to the Curriculum, Joint Master Degree in Borderology is a study in political philosophy located in the border zone between Russia and Norway, dealing with culture and cooperation in actual border zones, especially how local cross border cooperation may function as a bridge over traditional cultural and environmental differences. The study is exploring how cooperation in border zones may produce tools for solving conflicts both locally and on the global scene, thereby creating a new type of international dialogue mostly neglected within traditional state diplomacy.

On the level of content Joint Master programme in Borderology includes the following modules based on philosophical problematic such as "History and philosophy of science", "Values and borders", "Philosophy of language", "Phenomenology as experience of difference", "Ethics", "Critical thinking", "Professional identity", "Activity theory", "The Russian-Norwegian border region as a historic phenomenon", ""Border" phenomenon: thought, language, body" ets. The course Writing Subjectivity uses the student's personal and professional experience and practice as a basis for a deep reflection and text writing. The aim is to forward own self-understanding in perspective of" the other" in the border zone. The students will do this as a part of their philosophical development of knowledge and competence. The course Writing Subjectivity consists of six equal modules, each of these modules has its own content, learning outcome and literature. The central method in this development is called "essay as a critical investigating method". By using this method, the students exceed their first view on themselves and the other and develop new knowledge and self-understanding. This change is called "the Copernican inversion".

The course Borderology promotes the students own reflection and develop their personal and professional identity. The aim is to present the border between Norway and Russia and various other kinds of borders as part of changing global world, where fortifications, walls, barriers and fences and other means of identity control, primary directed against foreigners, can be turned into instruments of respect for the other in oneself. A change in our understanding of the border may affect the status of the border itself. The course Borderology consists of six modules also. Therefore, in the curriculum, there are so-called "traditional" for philosophical education courses and unique original courses reflects individual conceptual aspects of some researchers.

If the students fulfil the program successfully, they will:

- get a broader philosophical competence and training on scientific research methods, up to master degree level, build on his/her own practical knowledge and life's experience;
- be qualified to deliver a larger philosophical work in the field of borderology;
- be qualified to use their personal and professional experience and practice in their area of work as a basis for further development of knowledge and competence;
- be qualified to create arenas for cultural dialogue.

The master programme will qualify for jobs within cultural activity, business like activity, administration of fish, oil and petroleum in the Barents region, peace and administration like contact between Western Europe and Russian Institutions.

At this stage thirteen students attend the programme. This is the multicultural group of the students who are citizens of Norway and Russia and at the same time, most of them are also inhabitants of a border-zone. The students studying the programme are both the representatives of different age groups and different professions: teachers, managers, journalists, interpreters, lawyers. Being arranged at the seminars the dialogues of undergraduates based on the presentation of various written by them essays are certainly non-uniform and differ both the level and depth of comprehension of various phenomena. Nevertheless, the live voices being inspired with the process of generation of live thoughts, are heard in such dialogues.

From the point of view of the highest meanings laid in the education as the process of self-development, such continuous sense-making dialogues have to be the base of the personal development, extremely expanding the horizons. «For the students at Master in Borderology the dialogue is both the tool and the goal even though the development involves a dialectical overcoming of participatory thought. The student's activity is a kind of exploratory work where the goal is to develop something new, crossing the border in their self-understanding and build the ground for their philosophy» (Methi 2013: 133). In the border zone, in the conditions of intercultural communication when the students represent different cultures, different mentalities and different experience «the type of activity fluctuates between free and necessary activity, between open and closed dialogue, between a creation process and an exploring process» (ibid.).

It is precisely the principle of broadly conceived dialogue that became the foundation of the various intellectual constructions of the 'Northern dimension' in the border zone marked by the Norwegian city of Kirkenes and the Russian cities of Nikel and Murmansk, with the general tone of these construction being determined not so much by the intellectual interests of the individual participants, as by the common field of discussion that enabled different and even diametrically opposed approaches to raising problems to be brought together. One more thing has become clear in the context of this work of several years' duration, namely that our combined efforts have brought about an understanding, albeit a vague one at this stage, of ourselves through other people's images and ways of their organization, which amounts to having acquired some experience of understanding oneself through the Other and as the Other.

Immanent Voices (The Materials of the Qualitative Interviewing)

In the context of the contents of this article, it is important to give attention to some responses and estimates of the ideas of the programme, live voices of direct informants, which were heard during the high-quality interview elaborated by the authors of this article[5]. There are eleven people as the informants: the teachers, tutors and administrators of the programme representing the Russian party. Most of them have various appointments at Murmansk State Humanities University and at the same time participate in the implementation of teaching within Joint Master Degree Programme in Borderology. It is presented only some responses of the informants reflecting the specifics of perception of an idea of borderology and Master Degree Programme as its real implementation in the educational multicultural field. One the first questions that is asked informants was as follows: "What is attractive to you in the idea of borderology?". It means that the fact of the participation of this group of teachers in the implementation of Master Degree Programme assumes existence of interest in the philosophical and educational concept of the programme. The analysis of responses of the informants has allowed to say that attractiveness of idea of borderology is both the possibility of using a philosophical method approach in the course of teaching and thinking development of subjects of educational process which is observed during implementation of Master Degree Programme.

> **Informant 1**: (My answer refers to the Master's Degree project in general, i.e. more to its approach to teaching and studying than to the subject of Borderology.) It seemed attractive in several ways. Probably, the most important thing was the idea to theorize philosophically things that are normally taken to be very ordinary, so ordinary, in fact, that most people never stop to think of them as something that deserves philosophical reflection. It sounded like bringing philosophy back to where it began (and, maybe, where it ultimately belongs) – to the immediate practical concerns of 'normal' people who are not really that much into theory, but who sometimes wish they could make better sense of their lives. It showed that philosophy matters to everyone.

[5]The complex analysis of the results of this interview will be made by the undergraduate student of the Joint Master Degree Program Victoria Litvin within her Master Thesis. Approximate date of defending of thesis is November, 2016. We are thankful to Victoria for letting us use some of her material.

Informant 2: The idea of border is not new to me and this idea comes to be attractive in its comprehension. In this case the comprehension of idea of border is both in the consciousness of the teacher and in the consciousness of the undergraduate student. It seems to me that each assembly within this programme looks as a meeting or a dialogue of such reflections concerning the idea of border. Besides, during the process of such teaching naturally there is a dialogue, which modifies the borders of consciousness of both subjects of educational process, both the teacher and the student. Process of personal development and growth of the student is always attractive to the thinking teacher. And this personal growth is visible in the essays of undergraduates during 2 years of studying obviously.

Informant 3: Certainly, the idea of borderology looks attractive as it gives the chance to rethink a border phenomenon within an educational field of the Master Degree Programme. It is important that fact that the teaching is organized in the centre of a cross-border zone, students – most of them are inhabitants of a border-zone. It seems to me that here is the connection between the model of a real life in the conditions of trans-boundedness and ability or attempt to understand one's own existence in the light of various, not only especially geographical but also geopolitical boundaries.

Informant 4: As for myself as a teacher of the higher school the idea of Borderology is interesting through a prism of its implementation in the form of joint Master Degree Programme which is implemented by two different Universities which are on the territory of the different states. Teaching process as it seems to me allows to distinguish different borders between cultures (in this case it is obviously not said about outlined borders in other words geographical ones) and certain internal borders which often the person sets oneself.

Informant 5: Borderology is rather a project than idea for me. The project is attractive in view of a human factor: the team is magnificent, the mechanism of implementation of the project is built optimum, I see potential productivity. But the idea for me is still rather indistinct. In my opinion there is a danger in some indistinctness of a subject field, ambiguity of methodology. On the other hand it is attractive to me because it is connected with novelty of idea.

The second question was following: Do you consider that the borderology is something essentially new? Most of the informants reply positively to that question and talked about the basic novelty of idea of borderology at the level of a practical implementation of philosophical idea in the form of the Joint Master Degree Programme. Novelty is the formation of the new multicultural collective project, who is capable of carrying out their ideas in the course of dialogue of cultures. For this reason repeatedly by the informants, the "Borderology" is named the project, but not the programme or conceptual aim. Besides, essentially new, there are also some professional roles which are initially in the western educational system, such as, for example, a role of a tutor. According to the majority of respondents, the philosophical idea of borderology and a statement of such philosophical problem, developed in all variety of contexts and meanings, is not something essentially new. Here we present the most typical responses:

Informant 6: Certainly, it is not new idea but essentially new educational practice both for the Russian students and for the teachers. The method of writing essays as such is not unique and new for our programmes, but there is no doubt that the organization of the educational process on the basis of the student's reflections which are presented in the contexts of their essays is new. The knowledge is produced by the student from himself, on the basis of his reflections and after that it is already supported with this or that theoretical platform.

Informant 7: Two moments were essentially new as it seemed to me: the organization of the seminars on the basis of representation and analysis of the student's essays and work under the direction of the tutors. The institute of tutoring in itself is new to the Russian system, such appointment and according to such functionality has not been in our educational tradition. The tutor is only observing the process of discussion of the essays, occasionally directing it if this process goes out of the necessary course.

Informant 8: The role of the tutor was completely and totally new for me. It is usually very complicated just to look on the process of discussion at the seminar and let nature take its course. In Russian tradition, the teacher is as a rule very actively involved in the process of discussion taking the leading role in the process. It seems to me that in Borderology the role of the tutor is not of simply "outside observer" and more likely a role of "analyst" who is planting questions directed on expansion of a field of reflections.

Informant 9: I once again tried out as the organizer of the international joint program; I learned to be the tutor. In Russia, this position is absolutely new. Probably, I learned to listen to each student better. The Russian "classical" pedagogical scheme "from the theory – to practice" is still closer to me but having got an experience of this program I understood that other method of training can be quite effective.

Informant 10: The idea of border is not new neither in philosophical nor in historical scientific tradition but within participation in this programme I worked out a new course in English, approved it among Russian and Norwegian audience, received inspiring comments, new offers; new ideas were appeared. Generally this programme had helped me to realize my potential better.

Informant 11: From my point of view "Borderology" is the essentially new philosophical and educational project which has joined two Universities within one Master Degree Programme: Murmansk State Humanities University and University of Nordland in Bodø. Certainly, in the Russian and European institutions of higher education a number of the joint programmes are implemented within various disciplinary fields, but the implementation of the joint Master's programmes within especially philosophical perspective are obviously unique education, first of all it is owing to limit complexity of the meta language. Besides, the traditions of teaching philosophy in Russia and Norway are different absolutely. This programme has proved it once again. Russians pay more attention to study of history of philosophy, Norwegians – to develop the ability to place emphasis and to ask sense-making questions, being based on deep knowledge of philosophical texts.

The following question "What word is a key word of "Borderology" from your point of view?" correlates with responses to two previous questions. Informants defined the following words as the key words: "dialogue", "openness", "opportunities", "innovations", "cooperation", "culture" and "world". The interview did not assume the detailed explanation of some reason why this or other word was defined as a dominant therefore it is only possible to assume that dialogue, openness and cooperation are perceived by the informants as the basic principles of implementation of the programme; probably the world of different cultures is perceived as a source of boundless number of questions. The search for the answers to those questions is just provided by building of a number of innovative educational zones in the field of trans boundedness.

Conclusions

It may be said that the significance of such Master Degree Programmes today in a situation of increase of volumes of globalization and transformation of separate borders is incontrovertible. But it appears that full awareness of significance of the Norwegian-Russian cooperation reflected in such programmes will come to light not at once but in the long term of various vital strategy, which will be chosen by the graduates of the programmes for themselves, overcoming and expanding separate "borders". The idea of personal self-development of the student, filling with a deep meaning the content of studying and predetermining the uniqueness of its form, steadily leads to that the curriculum of the programme is renewed continuously, supplemented and extended by means of inclusion of separate substantially new elements in it. It is appeared new vectors in already developed courses; it is attracted new intellectual forces both Norway and Russia for the purpose of reading lecture courses. Nevertheless, at all general positive direction concerning prospects of development of this cooperation of institutions of higher education within the Joint Master Degree Programme, a number of the problem fields revealing all possible risks has been already obvious. Including the situation of lack of long-term sources of financing on implementation of Joint Master Degree Programme of the Russian partner, it is difficult to speak about full-fledged viability of any programme assuming symmetric responsibility of all members of university consortium. Thus it should be noticed that the preparatory stage focused on the implementation of the second admission for studying this programme confirms the attractiveness of the idea of borderology and such form of studying for future undergraduates. About forty people from the different countries of the world applied for a studying this programme, and it has been rather stable interest in such studying to the Russian students. Preliminary negotiations between coordinators of the programme from Russia and Norway indicate that within the second admission the substantial field of the programme will be significantly modified and expanded by means of inclusion in it the modules of economic and ecological plans. The focus will be shifted towards an ecological economy that will allow to be considered the interests of the regional development in the light of comprehension of phenomenon of border.

References

Burkel N, Tvorogova S, Shenderova S (2014) Innovations and changes in transnational education. European-Russian Joint Study Programmes. European Union 2014

ECA (2011) Joint programmes: too many cooks in the kitchen? The challenges for accreditation, recognition and transparency of joint programmes – a conference publication. http://ecahe.eu/w/images/c/cO/Eca-publication-joint-programmes-too-manu-cooks-inpth-kitchen-final.pdf. Accessed 21 June 2014

EUA, EMNEM: European Masters new evaluation methodology guidelines for higher education institutions (2006) – Guidelines for quality enhancement in European joint master programmes. P.9. http://www.eua.be/eua/jsp/en/upload/emnem_report.1147364824803.pdf. Accessed 17 July 2015

JOIMAN (2009) Good practice report for the administration and management of joint programmes. Annex 5: Glossary. http://www.joiman.eu/ProjectResults/PublicDeliverables. Accessed 17 July 2015

Methi JS (2013) Bakhtin and Vygotskij: polyphony, dialogue and the zone of proximal development. In: Sautkin A, Kopylov A (eds) Dialogue across borders: Borderology in the Bakhtinian perspective. Murmansk, pp. 125–134

Ryzhkova IV (2013) The Modern University in the Framework of Europe's "Northern Dimention": The Internationalization Phenomenon. Monograph. Murmansk, MSHU

Transnational education – Presentation of definitions and Code of Good practice based upon the results of Council of Europe/UNESCO Working Group on Transnational Education. http://www.ehea.info/Uploads/Seminars/trans_edu(1).pdf. Access: 17 July 2015

Part II
Challenges of Mental Borders: Challenge for Europe

The Power of Doubt

Boris Sokolov

Abstract In this article an attempt is made to suggest an explanation of some internal reasons for two interconnected tendencies in European culture. The first tendency is connected with a special way of creating frontiers in the European Union where they have now acquired a largely nominal character. The second tendency is a compulsive expansionism of European culture which, for long periods of time in the past, used to be realized in the shape of colonialism, and has now taken the form of globalization. Both these tendencies are connected with what the author defines as a "doubt imperative", which is not only a constant "efficient cause" of scientific discourse, but also a means of constituting everyday reality. The article discusses both the thematization of this "doubt imperative" in the history of ideas of the modern philosophical tradition from Descartes to Husserl, and its current understanding as a model for business and everyday life. The compulsive reproduction of the "doubt imperative" makes it possible to draw the conclusion that the "doubt imperative" belongs to the basic spheres of modern European culture and consciousness, i.e., that it is connected with the basic structure and style of the process of constituting European reality.

Keywords Border · Globalization · Archetype · European culture

Introduction. Statement of the Problem and Hypothesis

The initial impulse for writing this essay came from my reflections on two easily discernible tendencies in the European cultural tradition which appears to be intimately interrelated with each other. The first tendency which has taken a clearly defined shape in Europe in the past 40–50 years is connected with the way the European Union and, therefore, European culture constructs its borders. One can

B. Sokolov (✉)
St. Petersburg State University, Sankt-Peterburg, Russia
e-mail: sboris00@mail.ru

J. S. Methi et al. (eds.), *Borderology: Cross-disciplinary Insights from the Border Zone*, Springer Geography,
https://doi.org/10.1007/978-3-319-99392-8_5

hardly fail to notice a constant and persistent aspiration to create transparent borders, which ultimately leads to their actual abolition. This is true not only of the internal EU borders which now have a nominal character, and may even be said to be non-existent as "corporeal" objects, but also of the Union's external borders which are also increasingly called into question in the present situation of globalization when national or state associations are projected onto a globalization scale where they eventually lose their character. The European tradition and its reference points and models, whether economic, moral or those connected with values, are constructed and position themselves as "unlimited", making any border conventional and temporary.

The second problem, or the second tendency, is a compulsive expansionism of European culture which for quite a long time was realized in a colonial format, and has now assumed the form of globalization. Both phenomena are closely interrelated, as any form of expansionism is connected with a constant review of borders, with a logic of transcending any kind of limit. The European civilization has displayed throughout its history a constant striving for expansion, an urge to capture more and more new "territories", be it in the economy, in legal systems, in politics, in matters of daily life or in the sphere of value systems.

Numerous reasons have been suggested by way of explanation for such tendencies, both in the scientific sphere and in the sphere of political discourse, and even a mere enumeration of the various approaches and scenarios would undoubtedly occupy too much space. Many of them provide what would appear to be adequate explanations for the formation and existence of modern borders and for the connections between this phenomenon and the expansion of the European civilization. However, explanations for the process of the expansion of European culture, especially with reference to globalization, are often suggested on what may be called a "sectoral" basis, i.e., when clarifying the character of these tendencies, one is prone to identify one or more reasons pertaining to this or that particular sector of the whole of reality. But, most importantly, discussions of this kind do not normally take into account the question of what creates the "efficient causes" themselves which, in turn, create these tendencies. In order to identify what exactly constitutes in European culture its *causa causarum*, rather than the various particular causes pertaining to specific areas of cultural reality, it is necessary to turn to those areas of scientific knowledge which, while setting quite specific objectives for themselves, have always been guided by, and have always laid claim to, an understanding of the integrity and contextual interrelations of any subject under analysis. Philosophy, history, and cultural studies have always been considered to be such spheres of scientific knowledge. It is precisely in these areas of the humanities that it becomes possible to provide an explanation of the reasons for the compulsive striving of European culture to expansion and universalization connected with a constant abolition of any kind of restrictions, including frontiers. Therefore, this discussion, designed as it is to identify what exactly underlies the reasons for such tendencies, will mostly keep within the limits of the above-mentioned spheres, relying on the intellectual tools and methodology which may help to realize this project. What the author sees as such methodological reference points are, first and foremost, M. Heidegger's analytic of *Dasein* and E.Husserl's phenomenology, which makes it possible to clarify what it is that underlies the models and "archetypes" of a particular mode of "being-in-the world". This article also draws on a number of stimulating ideas concerning cultural traditions (including that of Western European) which were formulated by O. Spengler in his book *The Decline of the West (Der Untergang des Abendlandes)*, taking, of course, into account the changes that both European and world culture have undergone over the past hundred years.

Now to the hypothesis that makes it possible to explain the tendencies mentioned above. The European's mode of "being-in" and, consequently, the mode of structuring existence and its models in European culture, which is constantly aimed at expansion and doing away with all kinds of borders, are determined by two basic characteristics of this culture. The first essential feature of European consciousness (and the type of culture corresponding to this consciousness) is that of constituting the world as boundless. The second property of European consciousness, which is closely connected with the first one, is a special "style" of *Existieren* (to use M. Heidegger's term), a special "style" of existence which is infected with doubt, uncertainty and foundationlessness. It is precisely the fact of doubt being incorporated into what in the analytic of *Dasein* is defined as *Existieren* that not only ensures the perpetuation of expansionism, but is also responsible for what may be described as the power of the European cultural tradition.

The Doubt Imperative and the "Zone of the Indubitable" in European Culture

E.Husserl's explicitly formulated imperative contained in his famous *Cartesian Meditations,* which also happens to be the Cartesian imperative for building up a science, runs as follows: "The aim of the Meditations is a complete reforming of philosophy into a science grounded on an absolute foundation". "As we go on meditating in this manner and along this line, we beginning philosophers recognize that the Cartesian idea of a science (ultimately an all-embracing science) grounded on an absolute foundation, and absolutely justified, is none other than the idea that constantly furnishes guidance in all sciences and in their striving toward universality whatever may be the situation with respect to a de facto actualization of that idea" (Husserl 1982, p. 12).

This imperative of aspiring to an absolute foundation goes well beyond an imperative of scientific knowledge, it also embraces an aspiration belonging to the horizon of everyday experience, an aspiration to possess confidence in something undoubted, to rely on some kind of support which would guarantee the effectiveness of all our actions and the realization of all our plans. The purpose of scientific research is to achieve the zone of the indubitable which, like the firm ground we stand on, will ensure the creation of a sound body of scientific knowledge.

However, the purpose of any activity or object can never fail to influence this activity or object. It was a point discussed by Aristotle, for whom the purpose, or *telos*, was one of the reasons for the existence of any entity, i.e. something that forms this entity in its own image, since the purpose, or final end, serves as an ideal to which this entity must conform in a normative fashion. Thus, in the European scientific tradition a certain preferential zone emerges, the zone of the indubitable, to which every entity, not only scientific knowledge, aspires and on which it relies. European culture strives to reach the zone of the indubitable, it is intentionally aimed at finding its way to this zone. This means not only that the zone of the indubitable is a prime value of this culture, but also that this zone is formed and reached by a certain effort, proceeding from a situation which can never be described as indubitable. But the zone of the indubitable – as a purpose and as a value—is significant and important.

The zone of the indubitable does, indeed, make it possible, especially in extreme situations, to achieve positive results. Battles, revolutions, conquests, great religious shifts—though achieved at the price of heroism and self-abnegation—have always been the work of those who had no doubts as to the correctness of their ideas or the necessity of their acts. Even nowadays, the recent events involving Islamic extremism, which has, alas, only too often been successful and effective, demonstrate that in certain situations no room is left for doubt. Indeed, does the jihadist suicide bomber have any doubts as he prepares to sacrifice both himself and a number of innocent people for the sake of his inhuman ideas or beliefs? Of course, he does not. He doubts neither the correctness of his views, nor the need for human sacrifices, for spilling the blood of other people. It might also seem that it is in this aspiration which knows no doubts that his strength lies. He prepares to die without a shadow of a doubt. Even if he feels fear—it would be almost impossible to believe that he does not—this fear is overcome by the strength of his belief in the necessity of what he is about to do. It would be reasonable to assume, therefore, that you do not win thanks to your doubts, you only win when you have no doubts whatsoever. The jihadist wins because of his lack of doubt, as does the revolutionary who is out to destroy the old order, and as did the Christian in the first centuries AD, trying to abolish the powerful social machine of the Roman Empire and, with it, the entire classical antiquity.

Power has no doubts, it simply acts; this, of course, is applicable not only to European culture. The European "zone of the indubitable", however, is characterized by a feature which is specific to it and which clearly distinguishes it from the way such zones are created in other cultural traditions: in Europe, the indubitable is the purpose, the end, whereas the beginning, the starting point is something else. The starting point is doubt which is constantly overcome, only to be reinstated in its privileged position again, to be allowed to regain its initial effectiveness. The doubt imperative in European culture precedes the "zone of the indubitable", which shows it to be another cause, an equally effective cause also discussed by Aristotle, namely "the efficient, or moving, cause", i.e., "that which causes motion to start". The impact of doubt on the formation of modern European scientific knowledge, and, in fact, of any phenomenon of our reality, has been just as strong as, and even stronger than, that of the indubitable. This, too, is explicitly stated in the work with which we began our reflections, namely in Husserl's *Cartesian Meditations*: "When we turn to the content of the Meditations, so strange to us men of today, we find a regress to the philosophizing ego in a second and deeper sense: ego as subject of his pure cogitationes. The meditator executes this regress by the famous and very remarkable method of doubt. Aiming with radical consistency at absolute knowledge, he refuses to let himself accept anything as existent unless it is secured against every conceivable possibility of becoming doubtful". (Husserl 1982, p. 3)

Doubt is the genuine beginning, the efficient cause, of scientific knowledge, and it is responsible for building up the "zone of the indubitable" and the absolute. Doubt is an imperative or a principle which serves to cultivate in its own special way the soil of the "zone of the indubitable" on which scientific knowledge will grow, and which, in a broader sense, will produce each and every phenomenon of European reality. This kind of doubt imperative had been explicitly formulated long before it appeared in the works of the founder of phenomenology, namely at the very beginning of the modern European cultural tradition. The imperative, or principle, of doubt was given a thoroughly scientific thematization nearly four centuries ago in the writings of Descartes from whom Husserl borrowed it. This is how this principle is formulated by Cartesius in his *Principles of Philosophy*:

> 1. That whoever is searching after truth must, once in his life, doubt all things; insofar as this is possible.
>
> Since we were born as children and made various judgments, {some good and some bad}, concerning perceptible things before we had the complete use of our reason, we are diverted from a knowledge of the truth by many prejudices; and it seems that we cannot be freed from these unless we attempt, once in our life, to doubt all those things in which we find even the slightest suspicion of uncertainty.
>
> 2. That doubtful things must furthermore be held to be false.
>
> Indeed, it will also be useful to consider as false those things {in} which we {can imagine the least} doubt; so that we may the more clearly discover those things which are the most certain and most easy to know. (Descartes 1982, p. 3)
>
> 4. Why we can doubt perceptible things.
>
> And since we are now only concerned with seeking the truth, we shall begin by doubting whether any perceptible or imaginable things exist: first because we perceive that our senses sometimes err, and it is prudent never to place too much trust in whatever has even once deceived us; and next because every night in our dreams we seem to observe or to imagine innumerable things which are non-existent {elsewhere}; and to a man thus doubting, there appear no indications by which he can distinguish sleep from waking with certainty {and can know whether the thoughts which come in dreams are more false than the others}.
>
> 5. Why we can even doubt Mathematical demonstrations.
>
> We shall also doubt the remaining things which we formerly held to be most certain; even Mathematical demonstrations and even those principles which until now we thought to be known of themselves: 3 both because we see that sometimes some men have erred in such things, and have accepted as very certain and self-evident things which seemed to us false; and above all, because we have heard that there is a God who can do all things, and by whom we were created. For we do not know whether He chose to make us in such a way

> that we are always mistaken, even about those things which appear to us to be the best known of all; because it seems as possible that this could have occurred as that we should sometimes err, which, as we have already noted, does occur. And if we imagine ourselves to exist, not as a result of [an act of] a most powerful God, but either of ourselves, or of any other thing: the less powerful we consider the author of our origin, the more credible it will be that we are so imperfect that we are always mistaken. (Descartes 1982, p. 4–5)

Thus, according to Descartes, before starting to construct the building of science it is necessary to clean up "the territory of belief, delusions, unfounded opinions" etc., creating thereby a proper ground on which one can build up sound scientific knowledge. In other words, the starting point posited here for scientific knowledge and research is doubt, moreover, radical doubt. And it is only after "the machine of radical doubt" has tested a belief, a provision, a phenomenon of reality for soundness that it becomes possible to start building up scientific knowledge. It is true that Descartes primarily applies the principle of doubt to scientific research, to scientific knowledge, but this does not at all mean that the applicability of this principle is limited to the scientific sphere exclusively, in fact Descartes suggests that we should doubt much of what we normally believe in or what we have been led to believe in. The fact of the matter is that the task Descartes set to himself—and this is something that many writers on his philosophy have failed to appreciate—is by no means limited to proposing new principles and rules for the functioning of scientific knowledge; the task is to reform and re-encode the very "apparatus" which produces scientific knowledge (and therefore, knowledge in general), namely our consciousness, or mind, by means of introducing in a normative way a new mechanism (method) of its functioning. It is no coincidence that in his well-known *Rules for the Direction of the Mind* the very first rule proclaims the following: "The aim of our studies should be to direct the mind with a view to forming true and sound judgments about whatever comes before it" (Descartes 1985, p. 9). Modern European consciousness owes much of its "internal structure" to Descartes who was the first to set the parameters of the "scientific mind". Having "conquered" the domain of scientific knowledge, these parameters gradually and slowly became the parameters and imperatives of the European's "everyday consciousness" and, therefore, of the life world that creates this type of consciousness.

But now let us return to the doubt imperative which we found in Husserl, and three centuries before him, in Descartes—as the beginning and a constitutive principle of scientific knowledge. Both Descartes and Husserl formulate in this way two significant "figures" of European culture, namely the imperative of doubt and the zone of the indubitable which grows out of doubt and is therefore, in a sense, "secondary". The fact that there is a span of three centuries between the first thematization of the doubt imperative by Descartes and its "reiteration" by Husserl is highly significant: doubt does not lead to the absolutely indubitable, otherwise it would not have been repeated by Husserl after Descartes had already resorted to it. Besides, different variations, or "avatars", of this imperative had been a constant element in the history of European thought.

Thus, the zone of the indubitable is built up proceeding from, and thanks to, the doubt imperative. In this respect, no matter what Descartes or Husserl after him might have said about the indubitability of, say, the *cogito*, for them the beginning and, therefore, the really indubitable is doubt itself. Doubt is something that is not called into question, or, to use Husserl's terms, doubt is "apodictically indubitable", and therefore it is proceeding from doubt and with its help that one can start "clearing" the area of what cannot be doubted, which embraces, in fact, everything that one can think of, all that can be felt or imagined, etc. Doubt, therefore, does not only have a thoroughly "instrumental" significance, serving as a tool for preparing the basis on which a system can subsequently be built, it is also a genuine indubitable beginning which is the origin of any "gesture" of both European scientific thought and, by extension, of the movement of European existence constituting its own European life world.

It is no coincidence that the imperative and the principle of doubt are thematized by Descartes, i.e., at the very beginning of a cultural tradition which, following a certain scenario, constructs its borders and forms the basis and a model of modern globalized culture. Descartes did more than simply identify and formulate the central *topos* of this cultural tradition, namely the figure of the modern European subject, more than just thematize the tools of cognition and dominion which were given the name of methodology; as we have seen, he designated the truly indubitable, the *terra sacra* on which scientific knowledge is built up and all phenomena of the life world of modern European culture are constituted. Thus, Descartes is more than just an outstanding scientist, he is a genius who designated and thematized what was the most significant thing in modern European culture (F. Nietzsche was another thinker of genius who pronounced and demonstrated the constitutive principles of modern European culture). And—like F. Nietzsche—he believes that what he has discovered are generally valid imperatives (relevant to any cultural tradition), not just modern European imperatives.

Thus, "the gesture of doubt" has been recorded and given a most venerable role. Doubt is the only indubitable thing in the European pattern of knowing and constituting reality. And, therefore, doubt cannot be abolished, for in its case we are dealing with an "archetype" *sui generis* which can give any form to the indubitable, i.e., which is integrated into anything that is given and, primarily, into scientific conclusions and scientists' positions. At the level of the "rhetoric of the visible", though, everything may seem to be the other way round. For with Descartes, as well as with many other thinkers, doubt has to follow the scenario that Schiller reserved for Muley Hassan: "The Moor has done his duty, the Moor can go". Doubt has done its duty, it has cleared the territory for the indubitable, the evident, the apodictic, now it can be forgotten. However, this is not really the case, and the more recent history of European culture shows that "the doubt imperative" infects scholarship, in particular in the humanities, to such an extent that the gesture of doubt is repeated time and time again, while what was intended to be the final destination of the doubting scholar, namely the "zone of the indubitable", remains as unattainable as ever. The gesture of doubt is repeated, albeit in other forms, assuming other avatars.

It would be sufficient to quote only a few examples, the most significant ones, to demonstrate that "the worm of doubt" has never ceased "gnawing" at European thought. One and a half centuries after Descartes we can clearly discern the figure of doubt in I. Kant's critical philosophy which was, of course, intended to explore the limits and possibilities of the "tools" of cognition, but the gesture of positing the need for some preliminary research into the process of cognition, the gesture of calling it into question, is very much in the spirit of "the doubt imperative". In the European tradition one can find a whole group of thinkers who have to some extent used the "matrix" of doubt. Here belong, among others, all those destroyers of idols who, with various intentions and in their own manner and style, have devoted their efforts to the overthrow of the accepted and the indubitable. Among them are F. Bacon, M. Stirner, P. Feyerabend, Fr. Nietzsche and K. Marx, as well as a host of critics and subversives in the history of European thought in whose work the imperative of doubt has played a role which is not to be underestimated. Last but not least, E. Husserl, with whom our reflections began and who borrowed "the dubitative principle" of his philosophizing from R. Descartes, is not alien to doubt, either. As the *telos* of movement, "the zone of the indubitable" itself has seen its "rights" greatly curtailed, especially nowadays, in the format of knowledge found in the humanities which are guided by the idea of the plurality of truth, where it appears to be impossible "in principle" to reach the indubitable and the definitive.

Thus, doubt is a constant imperative which has been at work in modern European thought, and one which at present has little to do with the initial purpose of setting the "doubt machine" in motion, that of clearing the ground for the "zone of the indubitable". Moreover, the doubt imperative is gradually finding its way into our less-than-scientific everyday life, and can even become a model for successful business. For an illustration of this process of the imperative of doubt spreading far beyond the strictly scientific discourse, one can turn to a recently published book by Steven D'Souza and Diana Renner entitled *Not knowing. The art of turning uncertainty into opportunity*. The book itself is of little scientific interest, since the discussion of doubt on the conceptual level that it has to offer does not go any further than what can be found in the far earlier and more sound researches which have already been mentioned. In their *Not knowing. The art of turning uncertainty into opportunity* Steven D'Souza and Diana Renner elaborate in a largely journalistic manner on the not-so-novel thesis that the unresolved and the unknown can be productive, whereas complacency and lack of doubt can undermine progress and development. This thesis is supported by a plethora of biographical stories, and the book as a whole is written in a somewhat obsolete (but still popular) manner which was used as early as 1936 by Dale Carnegie in his best-selling *How to Win Friends and Influence People* which continues to inspire innumerable authors in this sphere. Therefore, the text (and the authors' activity as business coaches) is closer to Dale Carnegie's recipes for successful business than to scientific research. But this book is, of course, intended for a wide readership and is therefore written in a manner designed to appeal to as many readers as possible, promising them effective solutions to problems as varied as giving up smoking, learning to drive or becoming a successful fitness trainer. All this did not prevent the book being nominated for the 2014 Management Book of the Year (The Chartered Management Institute (CMI)). Maybe it was even helpful.

Now let us try to sum up what we have found out concerning the doubt imperative in modern European culture. This imperative was thematised in scientific discourse as a mechanism which was originally designed to ensure the indubitability of the indubitable. This imperative has been constantly reproduced not only in scientific discourse, but also in various spheres of reality in modern European culture, including the practices of everyday life and the patterns of doing business.

Doubt as a Style of *Existieren* of Modern European Culture

Thus, modern European culture constantly reproduces the doubt imperative, and this imperative can function both as a starting point, as a genuine foundation for the subsequent building up of a system (Descartes, Husserl, Kant, Nietzsche, etc.), and a pattern of doing business and achieving success in everyday life. Let us now ask ourselves: Why has this imperative in its different guises and avatars been compulsively reproduced throughout the history of modern Europe?

There may be several explanations because "the doubt imperative" is something fundamental. Various possible explanations seem to supplement each other and provide a complete understanding of this phenomenon. First of all, an explanation for a permanent "doubt imperative" in modern European culture may be suggested on the basis of S. Freud and K. Jung's psychoanalytic matrix. The constant and compulsive reproduction of the "doubt imperative" can be expected when "behind it" (or "through it") a source in the subconscious can be found, a certain archetype which constantly reproduces itself. However we should not try "to expand" the range of action of this "archetype", regarding it as a universal archetype of mankind. It is, most likely, a modern European archetype which, being an archetype, cannot be eliminated with the help of rational methodological efforts and tools, since in this case we are dealing with a layer of consciousness which, in fact, always remains outside the sphere of rational manipulations. The compulsive reproduction of the "doubt imperative" in the history of modern European thought, which we have cursorily traced, is evidence that an "archetype" is at work here, and it will necessarily "infect" the entire constituted reality. For an archetype, according to K. Jung, is a form into which all conscious processes are moulded, and therefore all these processes are "infected" with this archetype. In this respect, doubt belongs to the basic layer of European consciousness, determining the methods and patterns of its functioning and forming the kind of reality which can be described as the life world of the European.

Such functioning of the "doubt imperative" is possible only in the modern European type of consciousness which possesses a certain "optics" and structure. The "optics" and structure of modern European consciousness, let us use O. Spengler's idea here, is the "optics" of consciousness which constructs its world as infinite, i.e., which is capable of seeing infinity. It is a feature of the modern European view of reality that it sees this reality as infinite. It is characteristic of such "optics" of consciousness and sensibility that, on the one hand, the center of the Universe is located everywhere (Nicholas of Cusa), and on the other hand, this center constantly goes beyond itself. In an infinite Universe, it is impossible to fix that "one firm spot on which to stand" whence one could once and for all reach the "zone of the indubitable". In other words, the horizon of consciousness of European culture is infinite, as distinct, say, from the horizon of classical antiquity where the finitude of the universe served as a guarantee of definiteness and finality. For the classical mind, any entity—including the cosmos in general—had to be definite, for the lack of boundaries would have "plunged" both the individual entities, and the cosmos itself into chaos. The European "optics" of consciousness and sensibility, i.e., the "optics" of modern European culture, "can see" infinity, i.e., it constitutes the world as infinite. In this infinite world cognition is also infinite. And this infinity is "guaranteed" by doubt which is built into the forms of consciousness and sensibility. The European is in doubt, he is constantly in doubt, and even having reached the indubitable, he doubts this indubitable.

And the problem, of course, is not that the indubitable and the conclusive are unattainable; as Kant would have put it, they are beyond the limits of possible experience. The fact of the matter is that here we are dealing with some of the more fundamental properties of European mentality which—as forms and archetypes—are responsible for setting the parameters of our vision and of the way we constitute reality. Consequently, the "type" of person created by European culture, which in philosophical parlance is referred to as "subject", is not only someone who can see infinity, but also someone who doubts things, i.e., someone who is able only at the price of "titanic" efforts to reach the state of certainty and inner confidence, and even then for only a short time. Moreover, this confidence itself is not so much confidence as yet another avatar of doubt.

Doubt is constantly revived, or rather it never really goes away, since the European's knowledge is the knowledge of someone in doubt, it is incomplete by definition. Methodologically, this sphere now comprises all parts of reality. Even in the sphere of business and the social practices of coercion, the doubt imperative is winning more and more ground, proceeding methodically, as befits a principle inherent in the European tradition.

Thus, the imperative of doubt is connected with a general ontic disposition of European culture and consciousness, and is jointed with the central power points of this culture. Above all, with the "central point" of modern European culture which has always been designated as the modern European subject. It should be specified that this central element of all modern European culture—the subject—is essentially a foundationless entity, and therefore an extremely unstable one, it is someone who seeks to reassert himself (to receive an ontic alibi) in the course of constant self-affirmation. It is precisely because of his ontic instability and foundationlessness that the modern European subject constantly reproduces this instability thematically—doubting everyone and everything. The indubitable which is reached by means of "Doubt the cleaner" is always infected with doubt, and therefore the latter is always revived. Thus, doubt is something that may, for our purposes, be described as not even an *existentiale*, but as the "tonality" of the modern European *existentiales*, i.e., as a certain style of *existieren* which presupposes a process of constantly deleting the evident and restoring the figure of doubt. We are dealing with the intrinsic characteristics of the "phenomenon" which M. Heidegger defines as "Being-in": "Being-in is not a characteristic that is effected, or even just elicited, in a present-at-hand "in" something else that is present-in-hand; Being-in is rather an essential of Being of this entity itself" (Heidegger et al. 1962, p. 170). Doubt is a way, or mode, in which this Being-in can "realize" itself, i.e., it is a question of the style of *existieren*. For this reason this style will always be imposed on our every gesture, on our every action. In other words, the imperative of doubt is a style of our *existieren* which, precisely because it is a matter of the modes of our Being-in, will be constantly revealed and recorded in any phenomenon of ours, including the methods of drawing borders and, therefore, in the aspiration to constant expansion which has manifested itself in the form of colonialism and the processes of globalization.

Thus, the permanence of the imperative of doubt is a fundamental characteristic of modern European culture and its type of consciousness. It is precisely the fact that it belongs to this culture as a certain stylistic characteristic of its methods and models of constituting reality that has ensured a constant reproduction of this imperative throughout the history of modern European culture.

Conclusion

Let us recall that the starting point of our discussion was by no means the imperative of doubt, but rather the "less-than-serious" status of frontiers in the EU and the expansionism of modern European culture connected with this "understanding" of borders which originally took the form of colonialism and is nowadays proceeding in the format of the globalization processes. We addressed the problem and the status of doubt in modern European culture in order to understand the essence and the causes of the expansionism of European culture, as well as the nature of the constant aspiration, specific to the modern European cultural tradition, for an abolition of all boundaries and restrictions. Our analysis has revealed that there belongs to the basic spheres of modern European culture and consciousness what may be described as a "doubt imperative" connected with a certain structure and style of constituting reality. The "doubt imperative", which is permanently reproduced in scientific discourse and is gradually spreading to everyday life, is connected with a certain "optics" of modern European consciousness which constitutes reality as infinite, whereas the modern European subject who creates this reality is an unstable and foundationless entity which, therefore, permanently goes beyond himself. Doubt is the modern European subject's "style" of Being-in which predetermines both the methods of creating and then constantly abolishing limitations (including frontiers), and the expansionism of modern European culture.

This constant doubt which we have defined as the "doubt imperative" is a permanently running engine of development and expansion, and it is this engine that makes it possible to "capture" new "territories" and "spheres of influence". Therefore, in spite of the fact that in certain situations (such as war, conflict, etc.) doubt can be an obstacle preventing the realization of one's plans, it may be regarded as a European archetype of "Being-in"; the "doubt imperative" is an ontic structure which provides effectiveness and power. And therefore, the imperative of doubt is the power of European spirit, of European culture. "Within" European doubt, there lies hidden a powerful mechanism of constant movement and constant concern which enables European culture to be dominant on a global scale.

References

Descartes R (1982) Principles of philosophy. Translated, with Explanatory Notes, Miller VR, Miller RP. Kluwer Academic Publishers, Dordrecht, Boston, L.

Heidegger M, Being and Time (1962) Translated by John M, Edward R. Blackwell Publisher, Oxford, Cambridge

Husserl E (1982) Cartesian meditations. An introduction to phenomenology, 7th ed. Martinus Nijhoff Publishers, The Hague, Boston, L.

The Philosophical Writings of Descartes (1985) Vol. 1 of the philosophical writings of descartes, Translated by Cottingham J, Stoothoff R, Murdoch D. Cambridge, L., NY, New Rochelle, Melbourne, Sydney

Nature and Man: Crime and Punishment

Nadezhda Golik

Abstract This article presents a view of European culture as defined by three constants (science, education and art), and then proceeds to articulate the role of nature in culture. This view of the relation of culture and nature is then applied to the epochal event of the European Enlightenment. Due to the Enlightenment, which on the surface is a commitment to Reason, the system of values is changed. Nature becomes not a goal, but is reduced to a means. This crucial change can be seen from the perspective of crime and punishment, as a series of unintended ecological catastrophes. On this background there arises an idea of ecological aesthetics, which will correct the course of development and show an exit from the crises of culture.

Keywords European culture · Civilization · Nature · Sustainability of culture Enlightenment · Ecological catastrophe · Ecological aesthetics

The culture of a society is an indicator of its level of civilization. The culture of a society is a complex system. Like any complex system, a culture in its dynamics has the ability to develop, evolve and move to a different state. In some cases, this ability to move to a different state, can be classified as a decay (degradation) in culture and can be described as a cultural "crisis" (J. Baudrillard, F. Nietzsche, O. Spengler). The direction of cultural movement can influence both its internal and external factors, but in general, culture is a steady system in which its subsystems are in the process of mutual development and mutual influence.

If we take a closer look at the general direction of the development of culture, we will notice its many subsystems. Among the subsystems, we can distinguish science, education and art. They are the constants of what I will call the sustainability of culture in that they support its sustainability. It is however, because sometimes these three subsystems also functions negatively against cultural sustainability that

N. Golik (✉)
St. Petersburg State University, Saint Petersburg, Russia
e-mail: nagolik@yandex.ru

J. S. Methi et al. (eds.), *Borderology: Cross-disciplinary Insights from the Border Zone*, Springer Geography,
https://doi.org/10.1007/978-3-319-99392-8_6

Baudrillard, Nietzsche, and Spengler, not without justification, may describe the cultural dynamic of movement to a different state as a crisis.

Why are Baudrillard, Nietzsche, and Spengler only seeing parts of the whole picture, and for the most part are only able to focus on the negative developments of culture. One answer to this question is that they miss the importance of mechanical analogies. Therefore they are not able to understand the proper sustainability of culture. Using the analogy with the static stability of a mechanical system to understand cultural development, I will suggest that sustainability of culture is only possible if it is fulfilling at least two conditions: 1. hard links in some common intersection of all its three "fundamental pillars". That is, science, education, and art must have interactional links which mutually support each other, but without giving up its own special quality. 2. strength (or resistance to internal fracture) of each of fundamental pillars. That is, in the development of science, education and art, the must be some elements that cannot be corroded. If we look at culture, both of these conditions are fulfilled if value-moral coordinates permeates science, education and arts as a constant that provides its hardness and strength. Neglecting this relationship leads to a series of disasters: environmental, nuclear, anthropological. Symbolically it indicates the situation of "cancer" in a culture when the growth and reproduction of its own cells, is inhibiting the viability of its own health, starting to adjust the body to its dying—suicide.

The impact of science and technology on the values of culture and civilization in general and humans in particular, is therefore ambiguous. On the background of the historical development of the successes of scientific and technological achievements up to the present time, it has become obvious that they have dangerous social consequences. The twentieth century has not only confirmed Poincare's idea that science and ethics have different fields (ethics shows us the goal, science and technology offer the means to achieve them), but also have added to it: science and technology can bring real mortal danger to mankind, dooming it to the loss of the humanity in man.

Let us briefly consider the scientific and historical path that led to the present situation, which many modern scholars refer to by the symbolic phrase "crisis of culture". "Ecological catastrophe" is maybe another expression for this crisis. The idea of Husserl that our civilization since its birth, took the wrong way, received a clear confirmation when we look at the totality of the events of the twentieth century. But at first nobody noticed this phenomenon: Our thinking is focused on "a glaring symptom of" (A. V. Akhutin) and at the same time, the origins of the crisis remain not known, and it continues to grow steadily.

The philosophical search of origins, involves a complex work of immersion and comprehension of the concept of "nature", and this research is also likely to hide the truth about origins. This verdict is the conclusion of the old man of Western metaphysics, Martin Heidegger, according to him philosophy is such that it "hides some decisions according to the truth of things, if we recall that today the truth about creation in its totality remains in any case open to question,… and the essence of truth remains completely unresolved" (M. Heidegger).

Any attempt to clarify our ideas about nature may lead to an understanding of nature as a kind of regulative idea of the subject of all scientific disciplines. However, the process of clarification is not simple: The idea of nature is logically not very specific, and only intuitively believable. If we look back, classical philosophy was often on the verge of becoming a faulty natural philosophy, and the idea of nature was only presupposed without analysis, as if its meaning could be kept on silence. This was the case in the twentieth century's philosophy of science, epistemology, logic. In all these disciplines of scientific knowledge, the idea of nature was pushed it into the background.

However, in the late nineteenth century, there arose further problems which implied that this silence could not continue, the new developments in science and humanities made it clear that mind and nature were not to separate compartments. The Sciences and the Humanities cannot be diluted in two separate spheres, the spheres of nature and culture. Further development of the Sciences and the rapid emergence of so-called interdisciplinary fields of knowledge (e.g., bionics, bioethics, ecology, synergetic, ecological, aesthetics, etc.) by the end of the twentieth century, made it obvious that we need a different definition of our thinking concerning nature. According to Ahutin, we need a new definition of nature "for which the yardstick will be the mind, acting in the Humanities, particularly in cultural studies, historical hermeneutics, poetics, art history and philosophy." To sum up the situation: In turn, the growing separating of Mind and Culture implies the need to rethink the foundations of rationality—reason, defining "critical thinking as a scientific, moreover, as natural science".

The line of demarcation separating two types of mind and corresponding types of culture (as in Ch. Snow's famous work "the Two cultures") is threatening to break down any rational understanding of the subject. Bakhtin, Bibler, Akhutin consistently and clearly wrote about it. Akhutin, for example, shows that for some genuine historians to understand something "means to see something as a particular case of the action of a certain law under such conditions". But this for him means that "to understand means to establish the irreducibility of the phenomenon or event to the manifestation of the law. To understand something means to understand it in a unique independence of its essence. The art of this understanding is an art of distinguishing which have their own criteria of clarity, distinctness, rigorousness…."

As a perfect example of skill of discernment, Akhutin refers to Aristotle and his Physics, and his practice of going from that taken in General, to what is distinguished separately. And it should be emphasized that the discernment of Aristotle is logically related to his interpretation of "being", "essence", "physics" and is of fundamental importance to identify the universe, nature, which is the basis that for everything determines its way of being.

The knowledge of nature in the history of culture goes through several stages, whatever view we take (natural, technical, artistic, etc.). After the first stage as a kind of "initiation", a dedication to the external reality, and adapting to its requirements, comes the antique Space as a product of an eidetic mind.

The created world of Christian theology, too, only shows a limited understanding of nature. But the fundamental difference from the ancient understanding is the idea of a Creator who creates the world from non-being, nothingness. For example, St. Thomas Aquinas nature is a creative act of God who is the universal cause of all that occurs naturally. Since the world represents the perfection of the art of God—of his divine skill—the created world is for the "glory of God". The Cathedral and the Liturgy embody the Cosmos of Christian attitude, uniting every creature (unique and different) in the divine world order.

The beginning of the Renaissance in the seventeenth century prepares the "discovery" of nature. Two powerful figures of the Renaissance, Leonardo and Michelangelo are extremely similar in their intentions to the heroic enthusiasm of J. Bruno. The sculptor and the artist does not imitate nature, but instead discovers the "intentions of nature", freeing a block of material ("stone wild and full of power") from formlessness.

But if it is possible to talk about the origin of this intention, we should perhaps mention the famous ascent of Petrarch on Mont Ventoux in 1336. He made a detailed report in a letter to a friend describing all of his thoughts and experiences. There is a view according to which this ascension begins the whole Renaissance, so significant was this act. For example, Clarke lists it among those actions and traits of creativity of Petrarch, which allow to consider him like a man of a new type, initiating a new era: "…he was the first who climbed the mountain just to enjoy the view."

The next stage of "initiation" into the essence of nature is the attainment of the consciousness of the infinity of nature. This consciousness if not only produced by the scientific thinking in the New time, but by the development of Art.

If you try to relate the cosmology of antiquity and natural history to experimental art forms, it is possible to assert: It is in the mind of the artist that there is established a connection between the world and the cosmic order. The hidden appeal of the Cosmos and Nature to each other, their internal continuity is revealed in the act of creating works of art. One of the clearest examples of it is the poetry of M. V. Lomonosov.

Two main ideas permeate the content of Lomonosov's famous poem, "Morning meditation on God's Majesty" and also the poem called "Evening meditation on God's Majesty in the case of the great Northern lights". He gives us here a new idea of the worship of the power of the Creator of nature. This creator is creating thousands of unlimited worlds, each of which is filled with riddles and mysteries to solve what is called the human mind. The poet-scientist compares the universe of a "Burning forever Ocean" or abyss, with a universe filled with countless stars ("it Opened an abyss full of stars; the Stars are countless, the bottom of the abyss"). Still, the human mind—confronted with the unlimited—will be able to comprehend the laws of nature if it learns to ask "real" questions and tries to find the answers.

Lomonosov is extremely interesting for our topic, since in his work as lexicographer, he embodied the primary intention of intellectual activity that we also find in Francis Bacon. In his writings he is brighter than the other representatives of the New time's philosophy, expressing the essence of the mindset of this era by a new definition of nature as object of experience. While the fundamental idea or definition of nature is: nature is a different, unexpected, unforeseeable, like an ocean or a new continent, beyond the limits of human activity, still it might be known by means of a rigorous and difficult process of learning. It is in this context the concept of experience takes its new and modern meaning. Experience speaks here of a way of obedience, as a way to adapt the mind to nature itself, to its being. What we have here is not research based on tests and verifications, but an alternative way. This way at the same time prohibits the use of nature as a means, that was to become the prevailing intention in the social practices of the Enlightenment.

Here Nature becomes an object. In fact, nature is alienated. It not only becomes the object of aesthetic contemplation (Kant)—but at the same time the object of a physical experiment. Natura becomes the object of experience, for man to benefit from nature. In symbolic meaning Nature now is dead, and the man looks at it from the outside. As we know, that is what leads in the end to the emergence of serious environmental problems. The relation to nature is now based on scientific tests, and the technique uses nature to meet external needs. In the twentieth century, however, Ecology, ultimately, has to find the economic balance, allowing peoples to understand how to live in nature and not to destroy their own environment. However, the landscape becomes external and is perceived as external—that is important for its appearance and admiration. Critical opinion is both extremely concerned and extremely distant.

A famous slogan of Bacon's "knowledge is power" replaces the worship of God by the worship of Reason and then in practical experiences of Enlightenment, Reason is transformed into Usefulness. The latter becomes the basic concept of the Enlightenment and marks the birth of the well-known culture of scientific and technological progress with its inherent ethics of utilitarianism. It is this culture that, according to Ch. Snow, is fenced off from traditional humanitarian culture of the European West with an wall of misunderstanding. The optimistic universalism of the Enlightenment reveals the hard and cruel metamorphosis of classical relation of goals and means: nature is transformed into a giant storeroom of inventory of raw materials for production and energy, and together with nature and man becomes a tool. Utilitarian mind is assuming that a person is given to know about nature (including human nature, his biology, his physiology and psychology), what is better for nature and how to remake it for an adjusting to the culture. This misunderstanding is the result of the representations that are formed in the eighteenth century.

It is well known that harsh criticism of the Enlightenment project was conducted throughout the twentieth century by many authors, for instance by a book by the English scientist J. Grey with the symbolic name "The Funeral for Enlightenment ". But in spite of this, we did not arrange a "funeral for Enlightenment ". There still remains one fundamental problem posed by the Enlightenment which is unsolved until now. This is a problem of the Enlightenment itself. M. Foucault considers that the issue of what is Enlightenment, still remains the question, "on which the philosophy of the new age was not able to answer, but which it never managed to get rid of. Now, for two centuries, it repeats it in various forms. After all, from Hegel to Horkheimer or Habermas, including Nietzsche or Max Weber, there is almost no philosophy that directly or indirectly has not been faced with this question: what is the event we call the Enlightenment (Aufklärung), at least in some part behind what we currently think and do? And on the question of what is the contemporary situation in philosophy it can be answered: the philosophy which tried to answer the question rashly has been thrown up to it two centuries ago." This is important for the understanding of the current crises of culture - which no doubt is faced with a "gordian knot" of many problems.

Let us remember the most important details of the old discussion which led to the creation of the notion of Enlightenment.

The discussion was initiated by two entries published with a small interval in Germany in the same Berlin newspaper in 1784–85. The authors of these critical essays, Moses Mendelssohn, and Immanuel Kant were not familiar with the content of the text of the other. Still, the two texts were related, since both were replies to a remark made by the Berlin pastor Zolner about a certain confusion, "under the name of enlightenment". Mendelssohn agreed with Zolner on his reasons for insisting on the term enlightenment as a confusion, but insisted that the confusion did not only relate to the term enlightenment, but also to the terms culture and education: these three words, Mendelsohn writes, "has entered our language recently. As it seems, they belong only to literary language. The crowd hardly understand them…."

According to Mendelsohn, there was only one common denominator of the meaning of the three concepts: education, culture, enlightenment, namely a "manifestation of public life; the result of the diligence and efforts of people striving to improve their social status". This statement by Mendelsohn means that the people is becoming educated if the people achieves harmony of social status (thanks to the art and diligence) and always will remember about the purpose of man, and man's destiny. For Mendelsohn the latter is the measure and goal of all human aspirations and efforts, the starting point of all our thoughts, "if we don't want to lose ourselves."

Mendelsohn's formulation here raises the question about pseudo-culture, as we would say today. What is pseudo-culture? What are the signs of it? The signs mentioned by Mendelsohns more than 200 years ago, are external brilliance (or the gloss) and an appearance of sophistication which disappear when we subject them to the critical examination. Mendelssohn uses the term "varnish" (polish, lacquer). I want to recall that the varnish (lat. Politura—polish, finishing)—the material used for polishing wood products. Mendelssohn explains his use of the term politura in this way: only that nation reaches the true good whose politura (varnish) is the result of culture and education, and whose the external gloss and finesse are based on good quality". Noting the different relationship of culture and education between several different peoples, Germans, French, British, Chinese, he is looking back to the ancient Greek nations which held a high level in education and culture and finds that the best indicator of education and culture is language.

In a remark that may seem to anticipate von Humbolt, Meldelsohn starts a discussion not about terms and notions but about man's destiny. Seeing man's life as a citizen in the light of man's destiny, man's private and public life must correspond to certain constants, "measures" or ideals of art and education for the citizen to be a harmonious person. Mendelson points to the need for compliance with measures in any area. Otherwise, education and culture are "wrong", and will weaken the moral sense of people and lead to the selfishness, unbelief, superstition, anarchy, waste, debauchery, and slavery. I think we will hardly need comments to see the timelessness of all of this.

In Kant's article we encounter similar judgments, but in general the answer to the question "what is Enlightenment?" brings the idea to a completely different level: Kant attempts to show how to make man become enlightened.

Kant defines Enlightenment like this: "Enlightenment is man‘s exit from a state of infancy in which he is by his own fault. Infancy is the inability to use his reason without a guidance from someone else. Kant sees man's exit from his infancy as the implementation of his destiny and therefore as a moral duty.

To escape this infancy, man needs resolute determination and courage to use his reason without the guidance of someone else. Therefore Kant's main slogan of Enlightenment is "Sapere aude! - have courage to use your own mind.

I want to emphasize Kant's insistent repetition of the phrase "man's own fault" and the word "courage". From the first lines of his article, Kant indicates the moral role of personal self-consciousness. The person is responsible for the condition of his infancy, therefore to leave it and go into a state of maturity, a man can only do by himself, he must use his will, forcing himself into this particular effort, freed from any help of any guardian.

In other words, Kant defines "the exit from infancy" as the realization of his destiny, and, therefore, as man's moral duty, making man to be brave. To be courageous is not only to know, which in itself is associated with fearlessness, but to be able to make a bold move based on his own understanding. The slogan of the Enlightenment (Sapere aude!) like any slogan, is also a mandate to himself and to others: "to abandon education for themselves and especially for future generations", according to Kant, means "to violate and to trample upon the sacred rights of mankind." In practice, this means that to educate people in Kant's sense should prevent people from becoming actors directly involved in the process. This process is possible to the extent in which people will make their own choice, deciding to participate by voluntary participation.

Kant believes that man's transition to an age of adulthood is considered by the vast majority of people (and among them—the fair sex!) not only as difficult, but also as very dangerous because being in infancy is very convenient! "Infancy" designates the paralysis of the will, and conditions us to rely on an authority other than our own understanding: "If I have a book thinking instead of me, if I have a spiritual pastor whose conscience can replace mine, and my doctor is prescribing me a certain lifestyle, then I have nothing to trouble myself. I don't need to think, if I am able to pay; instead of me others will be engaged in this boring case". Thus enlightenment needs freedom, namely the freedom in all cases publicly to use your own mind ....

Kant is here connecting with the ideas presented by Mendelsohn about participating in two different spheres of human life, as a citizen and as a private person. Under "public use of one's own mind",—Kant writes,—"I understand that as a public person I am standing as a scholar in front of all readers". In other words, Kant is formulating the problem of transformation from man's infancy as a problem of reaching the level of a public use of reason. For him the scientist isn't free in the fulfillment of the duties of a citizen (taxes, discipline on service, etc.) if he remains in the immature private state. And to reach the public state for any person requires courage.

Today, we can add that a scientist is a person who is free in the space of leisure and reflections, and who in his public use of reason is strong enough to follow up his scientific ideas in spite of critics, prohibition and even destructive strategies from officials and bureaucrats. The problem of maturity in our media-society namely is that the executing of official power in the society still may be based on only the private use of reason.

Two hundred years later, Michel Foucault writes an article under the name "What Is Enlightenment?" and like the graduate student in the mode of "special reading" attempts to work out a new analysis of Kant's text about Enlightenment. Foucault establishes a connection between the main ideas of Kant's article and Kant's three "Critics". The necessity of Critic, as we know, results from the need to define conditions under which the use of reason is legitimate, establishing that we can know, what we should do and what we are allowed to hope. As Foucault is cleverly noticing: "The criticism is in some way the logbook of a reason entering the maturity in the age of Enlightenment and, in reverse, the Enlightenment is the age of Criticism". Using Foucault's point, one may say, that thanks to the "logbook" every moment of the events in the history contacts with past experience. This contact is necessary not only for our human reason to make a movement to the target with the smallest losses, but also for the definition of personal participation in this movement as a mature and public realization of reason. To realize a mature personality, it is in this critical perspective, necessary to set to oneself the task, to determine the "task" to himself so that it will determine your participation in modernity. I will add from myself: it is possible to perform the "task" of realizing one's maturity for oneself, but only on condition of a further and more personal "Copernican" revolution of the consciousness! What does it mean?

Such a "Copernican" revolution is made, when in the center of the world in place of the "I" with his "unique" passions, thoughts, desires, and whims, is appearing an understanding of being part of the common movement and responsibility for everything that happens. Then modernity is not a time modus in the epoch between past and future but the intention of consciousness, including a particular way of relationship to everything that is happening now, a free choice which not everyone may be capable of, a particular way to think and feel, a particular way to act and behave.

It is obvious that the Copernican revolution incorporates the movement from infancy to maturity and that it represents a development from private to public use of reason, but it is also something more. The personal Copernican revolution is impossible without the formation of a so-called culture of self-analysis which is constantly making "the Sisyphus's rise" (Camus A.). This means that the culture of self-analysis is connected with the ability of "moral reproach which the person has to deliver to himself". This analysis is not connected with "classification" of people, with value judgments in any way: one it is worse, others it is better. It means responsibility to go through with a moral and political task "to have courage in the work to go all the way what was seen in the soul and not to stop on democratic rituals of the social environment, an affiliation to any team" (Mamardashvily M.).

We see that Kant and Foucault, two main representatives of the intellectual European elite confess to the same general idea: the need of Europe to exit from a condition of "infancy". They are unanimous in their understanding of how it should, and how it should not be: no revolution is suitable for Enlightenment. I want to quote prophetical words of Kant: "By means of revolution it is possible to achieve, perhaps, an elimination of personal despotism and an oppression from a self-interest (cupidity) or a desire for power, but it is never impossible to carry out true reforms of thoughts by means of revolution; new prejudices, as well as old, will serve as main guidelines for a thoughtless crowd".

Two hundred years later Foucault supports Kant: "the historical ontology of ourselves has to turn away from all projects which lay claim to be radical and global. We know from the last experience that different way to think of the aspiration to escape system of the current events for the sake of universal programs of creating another society, another culture, another vision of the world, actually already led to renewal of the most dangerous traditions".

It should be emphasized that the Foucault's attention on Kant's text focuses on the fact that there exists an absolute value for the evaluation of the present and in retrospect for the past. Referring to the teachings of Kant, Foucault indicates that "rooted in the Enlightenment, philosophical questioning of a special kind, problematizes both the relevance of the present and history for the formation of oneself as an autonomous entity." For Foucault, it is important to underline that our inextricable link with the Enlightenment is not a commitment to its canonical principles and loyalty to its "philosophical ethos", but it is the urge to constantly recreate specific analysis of our historical existence ("critical ontology of ourselves"). The practical limits of such tests is that "we can become mature as we work on our own selves as free beings. Thus Enlightenment is presented as maturing a humanity which only begins to live, when they are using their own minds.

A crushing critique of the Enlightenment which began in the nineteenth century and continued in the twentieth believes that the modern Western civilization came to a deadlock because it was not able to fulfill the universal claims of the Enlightenment. This inability resulted in a crises which showed the failure of the rational justification of politics and morality. The shocking reality of modernity (ecological catastrophe, fundamentalism, ethnic and religious conflicts, the growth of social and ethnic intolerance, etc.) does not fit in with the optimistic picture of a future with "perpetual or eternal peace". But we should perhaps make one significant correction: Western civilization was exhausted by the project of the Enlightenment, but it had only in part embodied its full potential. Only one of its main tendencies—its orientation to the practical mind—was realized.

When Enlightenment is accused that its appeal to Reason as the last instance of its self-realization, had collapsed or failed, it is actually only a collapse of the practical mind. I remind you, in the German classics, reason opposed the mind as a higher capacity of the soul. The practical mind is occupied only with the relative and the finite aspects of the world, its business is find means to any given end, to learn to administer the earth. The Reason on the other hand, is capable of and must disclose the nature of the heaven, that is, the absolute, infinite, divine. In other words, the mind is focused only on the benefits from experience, the pragmatic, how to return a favor (remember the well known the article of Guy Bess "Usefulness as a basic concept of the Enlightenment").

Reason opens in man and mankind another dimension than the pragmatic orientation focusing on the cult of money, profit and adjustment to the circumstances.

Actually, Reason is that instance which creates from Homo Sapiens, the living being, the highest creation of the nature—namely the Person and determines his second, authentic, spiritual nature. The power of practical mind however, is doomed to the short-sightedness of a closing vision of strategic goals and possible metamorphoses in embodying the Idea of Reason in reality. Due to this fact, it can only be understood as a consequence of a "madness of self-importance" with which the rational thinking fearlessly makes attempts to prove its projects of reorganization. According to Hegel, all real is reasonable, but isn't rational.

The final result of the Enlightenment's orientation to rational (practical) Mind is well known to us now—the blossoming of consumer society connected with the "reset" of hierarchy of values: money victoriously took the place of the final purpose of man's life, having lost its status of means.

But the possibility of returning to the orientation to Reason still exists, it is embodied in the appearing of ecological aesthetics. By ecological aesthetics in this context, I refer not to Reason in one of its interpretations, but to an experience of seeing Reason as a whole. Ecological aesthetics seems to me to be the basic intention which may make the sense of Reason understandable to day for the dissolved practical mind—by uniting all practical interpretations for the benefit of our present environment. This intention can be summarized as follows: environmental aesthetics—it is always a definite awareness of the existential and infinite in nature asserting itself as an absolute value. In this way, it is an urge to reckon with, if it is carefully preserved and involved in the boundless horizon of theoretical research and practical actions.

In this meaning, the artist is a type of Eco-aesthetical person, because he brings into the world an ecological culture. He creates a world of "existential participation", by bringing Reason in contact with the natural substance of thing by means of his personal energy. And it is the artist who embodies the image of the person who is able to feel and understand how to coexist with nature as a equal being, taking care of it, and be responsible for it. For without the intimate dialogue with nature, without heightened sensitivity, responsiveness, and without a deep, sensitive imagination, the artist and his creation can hardly come into existence, and Reason and Imagination never meet.

I believe that the environmental aesthetics developing today may serve as one example embodiment of the universality of the "universal consciousness and liability arising out of the Reason" (Husserl). And as a form of human consciousness ecological aesthetics—if it is penetrating into people's minds—in reality gives the answer to the Kantian question of what is the Enlightenment.

Nature as Stoic

Viggo Rossvaer

Abstract This article is a presentation of borderology by relating the concept to three different senses of the word "nature"; (a) wild nature, (b) nature as dynamic activity and c) nature as Stoic. It is the third sense of "nature" that is my main concern. By "nature as Stoic" I am approaching the Stoic idea according to which Man and culture are embedded in Nature, due to Man's membership not only in his own, but also in the "city of Nature". My point is that Kant's engagement to establish cosmopolitan rights is based on a reinterpretation of the relation of man and nature by taking the Stoic idea of man as belonging to another city as his point of departure. To support my argument, the article launches an interpretation of Kant's idea of the Visitor, which makes the cosmopolitan right of visiting the other city politically concrete.

Keywords Cosmopolitan rights · Ecology · Cross border communication
Border zones · Subsidiarity · Green belt

1. What I will be discussing always was a challenge for borderology. Borders often appears as lines on a map. But from close up they reveal another feature. Borders are not only lines on a map, they are also as complex societies integrated in nature. It is this ecological dimension, to be found in Kant's justification of his idea of perennial peace, that is borderology's starting point.

I think man's basic consciousness of living in a natural setting is the clue to a peaceful future for man. It is the consciousness of his enclosure in nature, and not his continued domination of nature, that may lead to peace some time. But the Kantian conception of perennial peace is far from clear, even if it obvious is ecological.

The word ecology is often heard today. But what does the word mean? In this paper the perspective is ecological; and the word is made more precise by

V. Rossvaer (✉)
Nord University, Bodø, Norway
e-mail: viggo.rossvar@nord.no

J. S. Methi et al. (eds.), *Borderology: Cross-disciplinary Insights from the Border Zone*, Springer Geography,
https://doi.org/10.1007/978-3-319-99392-8_7

comparing it to three different senses of "nature" that must be distinguished from each other.

My discussion will start with nature as wild nature, then proceed to the second sense, nature understood as dynamic growth. With this second sense, we are approaching the idea of nature as an ecological dimension in the border zone. But it is only with the third sense, when nature is seen as breaking up and transcending the politically and socially defined borders of man, that the border zone has anything new to offer in the direction of integrating and embedding mankind.

Borderology is a peace study, but our Study Center, the Bakhtin/Kant Institute, is located far from Moscow, Bodoe and Oslo, up in the Northern Wilderness of the Barents Region. The location makes people's eyebrows go up, and they are asking: Is your peace study of borderology favoring wild nature?

Our study center is located on the border between Norway and Russia, close to the border with Finland. It is situated more precisely in Nikel, in the middle of the newly established Norwegian/Russian border zone, which includes both Kirkenes and Nikel as the main centers. We are near the borders, but far from everything else.

The Norwegian town of Kirkenes and the Russian town of Nikel have, like several other European cities, an agreement of mutual cooperation. This cooperation is based on an analogy with twin cities contracts between border cities farther south in Europe. Often a twin city contract is the result of a previous political separation, as for example when the drawing of new national borders have split one and the same city into two parts. But the interest to cooperate as twin cities may also be the result of a natural separation, as when border cities are separated by a common river.

But our two northern twin cities, Nikel and Kirkenes, are not separated in any way like that, they are separated by a real wilderness. This domination of wild nature over civilized land in this Northern Borderland, is making twin city cooperation take on a very special form: Connecting traffic takes 40 min of fast driving through wild nature.

The border zone is a relatively small and thin enclave, 60 km broad, on both sides of the Pasvik River, covering the 196 km on both sides of the Norwegian/Russian border. The border zone consists of 30 km of Norwegian, and 30 km of Russian land on both sides of the physical border. The border itself is open for crossing for the inhabitants of the zone. They may use special passports (grenseboerbevis) for border-crossing travel, but only inside the zone.

In the Common Declaration from 2010 concerning the border zone between Russia and Norway, the primary purposes of the Border Zone refers to for example better travel in the border areas, more cross border politics, but also broader cross border contacts between various professions on both sides of the border. This includes cooperation between for instance teachers and nurses from both countries. The Declaration therefore, is explicitly promoting the benefit of mutual learning in education and in the health sector.

The opening of the zone was an historic invitation to its inhabitants to share and develop the common educational and health practice in these sectors, and for each nationality to have the other nationality as its model for new democratic development of public freedom and control.

In our application for money from the Norwegian State (The Barents 2020 program), the borderology project wanted to establish an international master in Borderology as the first step in establishing an international workshop for nurses, teachers and other professions.

Communication at Master Level would provide the right kind of a climate for stimulating these various types of professions to learn from each other's practice. But also we wanted our students to study the border culture as a political phenomenon. We had the idea that the culture in the border zone was a cosmopolitan culture.

Admittedly, the decision to place the study center in the border zone, to locate it in the Childrens Art School, in the mining town of Nikel, far removed from the university cities of Bodoe and Murmansk, was a deliberate one. But it would be wrong to draw the conclusion from that, that we wanted to place the study center in the wilderness and even less that borderology is favoring wild Nature.

The location inside the zone reflects the fact that we wanted our students to enter into the border zone, to study a unique culture to find out if it was truly cosmopolitan. Only by participating in the practice of border life, could our program add new observations to the general study of peace research.

The location of the study center in Nikel, therefore, was based on the hermeneutical principle that life and culture in the border zone should be investigated from the inside, where the observer is also a participant.

The border zone is remarkable not only because it is new type of phenomenon, but mainly because it is unresearched. Thus, the point of placing the study center in the border zone, reflects, I believe, also the deep need for a new type of research on borders.

Nikel, where our center is located, is a town with approximately 15 000 inhabitants. One may say that Nikel as a town is a monument of the Soviet Union, straight in the middle of the Kola Wilderness. This distinguishes Nikel from Bodoe and Murmansk. But the unique position of Nikel to-day, is that it is a new type of center, a Twin City Center in a problematic Border Zone.

The focus on wild nature brings us back to the political problems of cooperation in this border region, problems so serious that it has led to many national and international protests. Today, people who visit Nikel, are not so much impressed by the wild nature there, as by the lack of wild nature. After driving through wild nature, coming either from Kirkenes or Murmansk, everybody will notice that fairly little is left of it in Nikel.

The visitor moving into Nikel from Norway or Murmansk will know the reason when they smell the smog from the Nikel Combinat. If you focus on wild nature, then you get a shock in the border zone.

Coming from Norway, you will notice the effect from the mining industry soon after leaving the Storskog border check point—on the road up along the Pasvik river. After some 45 min your will on certain days have to close all the windows in the car.

Then you see that the trees seems to have disappeared, and that there are no forests, only tree-skeletons instead of trees and undergrowth. And when you arrive and get out to fill petrol on the Shell station, you start to cough.

You get a feeling that you have not at all moved into wild nature, but into deeply intoxicated and spoiled nature. The border zone, therefore, also has a history about the disappearance of wild nature. Politically, our border zone may be seen also as an experiment to solve a problem of destruction of nature, problems which neither Russia and Norway were able to tackle by themselves.

To sum up: Nature in the sense of wild nature may seem attractive as a picture of the best resources of the border zone, but the reality is different. In this wilderness we hear a cry from nature. The Nikel plant shows the problem: What can people do when nature dies?

The border zone is special because it demonstrates overlooked cosmopolitan resources that challenges the passivity of nations when confronted with the global killing of nature by man. The focus of borderology here, does not lie with preserving the wild nature, but with a wish to activate an international rescue operation when nature is seen to die.

2. Let us turn to the Green Belt, a term related to ecology, but also to geopolitics. The Green Belt of Europe transforms the perspective of wild nature into a new and more culturally oriented ecological perspective. Cooperation concerning the Green Belt represents the beginning of an international responsible politics in border areas: The unintentional destruction of nature by man can only can be prevented by the peaceful cooperation between nations, for instance by establishing multi-national border zones along the Green Belt.

The Green Belt is of interest for borderology, since it constitutes a long border belt on the European map. The Belt is thin as a curtain, but also very long, stretching throughout the whole of Europe. It marks an enormous path, demonstrating a Green dynamics. But the Green Belt was not originally brought into existence for ecological reasons. The Green Belt has a political history, starting after the Second War and the division of Europe into East and West.

The Iron Curtain is a Western concept, and little used in for instance Russia. Much more understandable in this context is the term "Berlin Wall", which is the name of the most spectacular part of the Iron Curtain, that short part of it which cold war politics mostly was centered on.

It was only after the removal of the Berlin Wall that the Green Belt, stretching from Kirkenes in the North to the Adriatic Sea and Odessa on the shores of the Black Sea, came to the surface as a concept. The Belt came quite as a surprise to biologists and naturalists all over Europe. While the Iron Curtain insisted on separation and control, the existence of the Green Belt showed the effect a totally different and free dynamics.

The expression "Green Belt" originally comes from city-planning, more precisely planning for the future to retain wild or undeveloped land surrounding urban areas. As a concept of city-planning the Green Belt has associations with setting up a border. In the first place a fence around an urban area, allowing wild forms of life to be established.

In speaking of The European Green Belt, one is speaking of a Europe defining itself along the East-West axis by a border belt connecting border areas of 24 different nations. In its wake there are already 52 national parks and 3200 protected areas in 24 countries.

Whereas the Green Belt policy originally referred to the policy of city-planning, the European Green Belt can be seen as a policy to transform the strategies of European power politics by introducing a different strategy for borders.

The Curtain that for many years had prevented men and women from moving from East to West or from West to East to in Europe, had during these year served as a very effective protection for many natural kinds of species, various forms of wild life, plants, animals, birds and other kinds of organisms.

The free and self-going dynamics of the life forms—when they were left to themselves—came as a symbolic proof that nature may be of assistance in producing an alternative and more peaceful future.

The expression "border dynamics" is used by Paul Eric Aspholm in his biologically oriented research, but is also used as a social description of border life, in Finnish migration theory. It dynamics shows a sense of nature that manifests its independence relative to human manipulations.

Nature is not only something wild and uncivilized, but also something dynamic. The sense of the word "nature" introduced by the development of the Green Belt reveals that nature let in peace takes its own self-protecting course. Let us call this the ecological sense of the word "nature".

For borderology there is a lesson to be learned from seeing political planning in analogy with the Green Belt, namely the power from below. The relaxation of control on both sides of a border may inspire a development of political power from below, as a characteristic dormant in the cosmopolitan culture of the border regions.

In this way, the analogy with the Green Belt confirms a half-forgotten Platonic category for state-justice, characterized by the old name of power from below taken from Plato's Republic, namely *subsidiarity*.

It the border regions of Europe today, it might be important to have this in mind, if one is going to understand border cultures and their relations to the bordering nations. Border areas have, if we rely on the analogy with the Green Belt, a surprising potential as cosmopolitan workshops for future international development.

I want to conclude the discussion of this potential in the border zones by referring to an interview with a teacher in Nikel. The students had pointed out to him that the communication in Nikel between Norwegians and Russians seemed to be opening up, creating new mutual respect and communication by just being together. Communication was opening up, even if people did not necessarily understand the spoken and written language of the person they had contact with. For them this phenomenon was a great paradox.

The students suggested that the communication was a result of Norwegian and Russians visiting each other more often, increasing business and trade, after the establishment of the border zone. The surprise for them was that the "non-regulated" communication was "democracy-building" in that it created mutual respect and trust. It was like bringing back the trust established during the time of the Pomor trade.

But they did not register any new social and democratic rules. And they asked their teacher how can this be a development of democracy when there are established no new democratic rules? The teacher at first had no answer.

According to his students, the contact due to the border zone invited its inhabitant to more intensive social contact between Norwegians and Russians. But how could a description of this contact prove anything about democracy? The increased contact was not reflected in any new explicit rules, the more intensified contact only seemed to make the social situation more chaotic. So the students said: Sorry we cannot find any proof in the form of democratic rules of a development of democracy in the zone.

We only registered non-regulated and risky social communication in the form of examples such as this: When Norwegians meet Russian at the dinner or lunch tables, they learn not only about new types of food, they also experience new cultural facts. Norwegians are for instance surprised at the use of meals for establishing personal relations, not hurrying at the table, but taking the time to bring all participant in a good and reflective mood. Many Norwegians have at a table in Nikel made a new and lasting experience; that eating together may produce friendship. Unexpected meetings often produced a feeling of deep social togetherness they had never felt before.

Taken in by this report, the teacher suddenly understood what the students were driving at, and his answer was this: Yes you see, the only way a democracy can exist, is as a practice established from below. Democracy cannot be established by rules, torn up from its roots and implanted somewhere else where these roots are not present. Therefore, it is wrong to think of democracy as existing by virtue of rules. It is thriving on something more basic, namely certain events and practices, concretizing freedom and respect ahead of rules.

This point the students understood: To observe the development of democracy, they had to give up to find social rules as confirmations, and rely on their own observation about a practice that is prior to any application of rules. They had to look for something dynamic in the behavior connecting people ahead of rules and institutions.

The students concluded that the social contact between different nationalities in the zone, was the heart of every real democracy. "Democracies" who have their legitimation in certain rules only, often are false export-democracies, imitating justice, but not based on the concrete interest to develop the citizens' *own* form of life. Democractic development cannot be thought of as imitation of an established normality imported from abroad. A living democracy does not exist by virtue of its rules, but by virtue of the practicing of its rules, which is a totally different and unregulated social dynamics.

3. The dynamic character of nature in the Green Belt leads us into the Stoic vision of the whole of nature as a dynamic cosmic context. But the Stoics took the dynamic relation between man and nature much more seriously that modern philosophers do in Europe today.

The Stoics provided us with an important sense of nature that in modernity to a large extent is forgotten. They saw man as a cosmic creature, but often living under social and political conditions that separated man from the Cosmos.

The Green Belt inspires a dynamic sense of nature, but according to the Stoic philosophers, man and his world are not originally separated, but basically parts of a bigger whole, an Unlimited Nature with dynamic borders.

This means that we have a third sense of nature to discuss, in addition to the wild nature and the dynamism of a nature liberated from domination by man.

According to the Stoic tradition, the natural dynamics of Cosmos, is the mother of all civilizations. It implies that man should listen to nature and even try to live in accordance with nature to understand himself.

The Stoic philosophers with their cosmopolitan view of state and society were opposed to the Platonic and Aristotelian philosophical tradition. They denied the value of regarded the city-state as defined by a geographically area. According to them freedom cannot be promoted by strong borders.

It is my thesis that the peace philosophy of Immanuel Kant is colored by the memory of the Stoic vision. And I will argue that it also gives Kant's idea of the perennial peace a distinctive flavor. Though never explicitly admitted by Kant, the Stoic attitude to nature is always there in the background in his "Zum ewigen Frieden", reminding the reader that culture is always embedded in nature.

In Kant, the idea of man as a cosmic creature centers on the idea of a visitor. The word "visitor" may come as a surprise. Why does Kant represent cosmopolitan rights by speaking only of *visitors*? But the background for Kant idea of the visitor, is a special version of how the Stoic idea of living in accordance with nature, should be practiced politically.

This visitor implements Kant's Stoic idea that Man's attempt to establish future social harmony cannot start with the regulation of human life inside the city's own borders. Instead, it must start with breaking up the borders. To neglect the primacy of Nature, is actually to neglect the inner cosmos in Man. The inner cosmos is already at work before any social regulation, and tuned into Nature. It means that men are communicating with Nature, like birds on a string.

For the Stoics, true experience of the cosmic harmony cannot be developed by prisoners of the city states. Such prisoners cannot get to know themselves fully. As cosmic creatures, however, they know of a different life. In virtue of their inner cosmos, everybody inside the city, is connected with unknown borders and events going on outside the limits of their cities. Due to man's nature, there is always a second city, better able to realize one's humanity, that the actual one.

To develop knowledge of these borders, man has to leave the city behind. By crossing the borders of his own city, he may be able to decode the meaning of nature and fight the borders that restricts his true freedom. Only then, outside the established limits, will he transcend his previous knowledge and realize his full potential as a human being. He will then discover the sense of the Unlimited—which keeps everybody together in the bigger whole of Cosmos.

The life in the local city must therefore be adjusted by the life in the true city of Nature, whose borders always can be given a universal moral justification.

This third sense of nature was important in the political philosophy of antiquity, its paradoxical sense of nature and citizenship was challenging for the articulation of the traditional city-state as the ideal of true democracy.

But more important for us, this Stoic sense of "nature" also defines a sense of border that is essential for the political importance of the border zone. In borderology the border zone, or more generally a border area, is regarded as having a form of political and social potential for developing new social and political meaning, challenging the sovereignty of the bordering nations on both sides.

Kant obviously thought so too. The visitor is Kant's name for a visionary figure who is advocating a fight for man's rights, by inducing individuals to take peaceful action against the modern state.

Kant's main argument is that the visitors' right is a new type of human rights: in the sense that every single individual has the right to react against dehumanization coming from the building of walls around national states.

Like the Stoics Kant is calling this human right cosmopolitan. And like the Stoics, such cosmopolitan rights concerns man's relation to the whole world and can only be discovered and developed by the inhabitants of one national state, through their operating as visitors seeing contact with inhabitants from another national state, through a sociality articulated by cross border contacts.

The border zone is an example of how history by accident builds up political institutions which will profit politically from the Stoic view of living in accordance with Nature.

From the Kantian point of view, the Stoic sense of nature shows that the border zone between Russia and Norway, can become loaded with a meaning that foreign ministers who are declaring such a zone for opened, are not likely to accept or even understand.

4. Based on the analysis of the Stoic idea of living in accordance with nature, one may wonder if foreign ministers Sergej Lavrov and Jonas Gahr Støre realized what they were doing - and what kind of local feelings they were touching on—when they opened the border zone in 2010.

In contrast to the two foreign ministers, people in the zone could foresee that the new passports would let loose old, neglected and even forbidden feelings of cross border contact in the North. The border zone passport does not only point to a new future, it is also an awakening of the past; more precisely of the big historic loss of extensive cross border contact established by the Pomor trade.

At the time the border zone was opened, I did some interviews with a number of public figures, among them some local spokesmen for the ethnic Finnish inhabitants in the area.

My main informant on the Norwegian side told me that due to the opening of the border zone, the people of Sør-Varanger started to remember old promises about local Norwegian-Finnish projects. They would for instance like to ask ministers Lavrov and Gahr Støre when the road across the border into Finland should be completed. Another person added: "There is still some 50 ms of unbuilt road separating the Finnish and the Norwegian road systems. If the 50 meters of unbuilt road is completed, that will make it much more easy to visit our relatives, and also to do business with Finland".

The problem with a connection to the Finnish road system has always been the argument it would be to invite an invasion from the East - not from Finland, but from Russia. Because of this possible threat, building the road to Finland is constantly postponed. The fifty meter of sump and wood, separating the Norwegian and the Finnish road system is serving as a mental defence against a possible Russian invasion.

With modern technology the danger of an invasion seems to day to be much greater than before, since the flat land of Finnmark seems to be indefensible from a military point of view. Therefore, the people living in Finnmark are feeling that they are treated secondary citizens. According to the national defence policy, people living close to the Russian border also constitute a problem and not only a resource.

Last year, however, in the municipality of Nordreisa to the north of Tromsoe, and close to the Finnish border, the border population demonstrated new ideas to increase cross border activity. Nordreisa is known for a population where many are of Finnish origin, and the municipality council wanted to celebrate the 100 year birthday of Finland as an independent nation. They decided to give away one of its mountain tops to Finland. The flat northern land with mountains and lakes which Norway does find worthy of defence might still be worth something seen from the Finnish point of view.

The municipality council wanted to give away the top of the local mountain Halti to Finland as a Congratulation for being 100 years old. Because of its height, the top of the Halti is the symbol for the Nordreisa municipality, and its big municipal hall is also called Halti, named after the top.

If the top of Halti became a part of Finland, perhaps this would perhaps give a boost to Finnish pride, they thought. The gift might perhaps also be used to strengthen the cultural relations to Finland. It would symbolize Finnish independence and Norwegian brotherhood. The idea of this gift also had a very practical context, related to common national interests in the growing tourist business in the North, connecting Finland and Norway.

In Norwegian national papers the gift was not taken seriously. But the offer got a positive response in Finland. The positive response maybe needs some explanation: In Finland this kind of gift was, because of the geography, something that couldn't be taken as a joke.

Geographically Norway and Finland are bordering on each other in the Halti area, the two countries are connected to each other by the dominating Halti mountain. But most of this mountainous area lies already in Finland. Therefore, for Norway this gift would mean little geographically, since the present border passes close to the very top of Halti.

But regarded from the Finnish side, the top of Finland's biggest and highest mountain lies in Norway—only some 40–50 m away from Finnish border. If one draws the border only 50 m to the northwest, giving away the top, the whole southern side of the mountain side would become part of a Finnish landscape.

Suddenly, at last the biggest mountain of Finland would have its peak on Finnish, and not on foreign soil.

To give the top of Halti to Finland, for the Nordreisa population, would not be much of a loss. Finland would get some square meters high altitude land, but also a big new tourist attraction. Also business in Nordreisa would get something in return, for many international tourists would come visiting Finland's highest mountain through Norway.

So what happened? The vote from the citizens in the municipality of Nordreisa at last ended in the Norwegian Parliament.

In the Norwegian Parliament, however, no one was willing to give way "one single centimeter of Norwegian territory". So the top of Halti mountain was not given away as a gift celebrating Finnish independence. It is still Norwegian.

5. My interviews from the border zone reveal an unexpected, but strong Stoic sensibility among ordinary people in the far north of Europe. I will present the main results of my investigation in the form of a conjectural conclusion: *Nature is Stoic.*

My material indicates that people speak of wild nature and landscapes in nature as offering a dynamics that breaks up the well-defined and often enclosing borders of their social world, and they find some emotional freedom in that. To put it short, I find that the reference to the nature confirms an emotional landscape underlying people's established and rule-governed relations to each other.

In such statements I see a confirmation that the social world is embedded in a nature, and therefore not easily limited by borders, in a sense that comes close to old Stoic insights.

Now, according to the familiar Platonic and Aristotelean construction of the world, absolutely nothing that exists can be without limits; every x, to exist fully, must have a limit; if x is without limits, x ceases to be. The unlimited in nature simply cannot exist.

Because of the dominance of the Platonic/Aristotelian tradition in Europe, this denial of the unlimited has become our main philosophical myth: It is only by means of introducing limits to things that is it possible to distinguish x from other things and explicate what it is. So the unlimited is nowhere to be found.

In Kant's writing about the perennial peace, however, he goes back to Stoic ideas, even if he is not much explicit about it. In his work on the perennial peace, for instance, it is easy to misunderstand his ideas about the "assistance of nature" in producing peace. These statements are often understood as teleological dreams only, even if they are basically part of his Stoic building materials.

I will try to explicate this Stoic sentiment in Kant by dressing up Kant's figure of the visitor as a person introducing the unlimited into politics. Starting up, the true philosopher must become a visitor. His aim is to develop the idea that the national state is embedded in a wider Nature with borders of a different kind.

The Kantian visitor is not primarily a travelling person applying for visa, but a person with a clear perception of his own subjectivity as being dependent upon an unlimited nature. He sees nature as providing man with a new horizon, writing letters of protest against the borders set by the national authorities. Accordingly, the drive and motivation of the Kantian visitor is a drive, not to become a tourist, but to build a second, more just city in a world lacking peace.

The perception of the unlimited has woken him up from the dogmatic slumber of his normal perceptions, and he now sees a new truth in a dynamics in nature, tearing things away from the system of normal perceptions and placing national borders in the middle of the context of the unlimited.

As often as not the visitor is not getting any permission by the national authorities to travel. In his attempts to cross the border, the visitor is stopped short. But the main problem with the visitor's permission to travel, is that his new vision breaks with the dominant Platonic/Aristotelian intellectual tradition in political philosophy and modern border theory.

The triumph of the Aristotelian/Platonic worldview is illustrated by the order of the modern state, which is organized from the top and ordered by public rules. But it creates tragedies at the borders. Accordingly, the borders of the old Platonic Greek city states still lurks in the background for the modern rationality, influences our perceptions, drawing bordering lines separating Man and nature.

And worse, the opposition to the visitor in Western history is not only theoretical, but moral. The basic Aristotelian argument against the Stoics is that the Stoic vision lacks the idea of moral self-realization: The Stoic permission of the Unlimited as a possible experience, invites moral chaos.

For an Aristotelian philosopher, to exist as a moral person is to be able to develop your potentialities, in a dynamic development based on clearly defined internal limits. To lack such limits and practice a policy of a cosmopolitan sharing is therefore the same as having no meaningful project for possible future self-realization.

Persons who are not able to realize the potential of their limits, cannot be fully masters of themselves according to the Aristotelian concept. They do not develop strong limits in their inner constitutions, and are liable to give up their resistance against strangers and join in with the joint cries from their unknown neighbors outside the national borders, who are wanting to come in.

6. What is the consequence for man if nature is Stoic? One conclusion is that it answers Kant's basic question "What is man?" For Kant man is basically a being who has to cultivate his consciousness of the unlimited, simply in order to keep up its fight to remain human. This consciousness includes feelings that are simple enough to register.

To present you with such feelings, I will take you out of the seminar room and into the nature outside our lecture rooms in Svanhovd. Our project of borderology, started many years ago, discussing Kant and Dostoevsky in general, has led us to concrete questions about feelings.

Marching out of the seminar, we are setting our feet on the soil which is part of the Green Belt. The landscape we have under our feet is sloping down to the banks of the Pasvik river and our eyes are resting on the Russian hills over on the other side. But be careful, here, in this most peaceful place along the Pasvik river, there is no connection over to the other side; throwing a stone over the middle of the river, will cost you a fine of 800 Euro.

We are sniffing at the not so fresh air, because the wind to-day is carrying some part of the intoxicated Nikel air over the Pasvik river, but our group have the time for a last reflection.

The beautiful river landscape on both sides makes up a zone with a natural water dynamics that in spite of being divided by the border, makes the landscape feel unitary and powerful. You may see the flow of the river as the originator of its own time. And you are part of it; you feel embedded in a nature which seems to enclose you totally, without any limits imposed by national states.

The feeling of being a part of an unlimited nature in this sense, is produced by a mild feeling of being at home in nature, a feeling that nature, more precisely, is unreplacable for your existence here and now. The feeling that nature as unreplacable completes your identity in a mild realization of seeing nature as one.

I think this is an important feeling. Feelings like this can motivate man to work for peace, since they bring up a feeling of cosmopolitan identity that transcends every national border.

A cosmopolitan right is the right of an individual to set his own individual rights up against the jurisdiction of a whole state. A cosmopolitan right is universal, which means that the right is valid for all individuals and that such rights extends beyond any particular state. But the problem is: How can this cosmopolitan universality be justified? What makes it possible to defend such rights?

It would seem that the universality of the cosmopolitan rights is justified by reference to some basic facts about Man. Everything points in the direction that the experience validating such a universal right is a feeling that man is embedded in Nature. While the moral law in Kant's moral theory is based on a fact of Reason, the cosmopolitan rights are based on a fact of Nature, the feeling that nature is unreplaceable.

When the question: "What is Man?" comes up in Kant's philosophy, Kant sides with the Stoics. Now I want to conclude that Man is a being who feels his existence in Nature as unreplaceable.

It is this common feeling that substantiates Kant's ideas of the rights of the visitor and his reflections concerning perennial peace.

I am convinced that it is this feeling, more than anything else, which is the reason that Kant calls the peace he is hoping for, for *perennial* peace. It is the feeling that you are involved in an unlimited nature, and that this experience gives you a passport as a citizen of the world.

This feeling goes a long way beyond any experience of participating in the biological dynamism of nature, since it connects the whole landscape with a feeling of who you are. I think that it is the feeling that nature is unreplaceable for your own human identity, that gives the Kantian visitor his drive. It completes a sense of being embedded in nature by a vision of the infinite world community as your natural home.

What Role Plays Intuition in Mathematics and Science? On the Borders Between Several Conceptions of What It Means to Intuit

Johan Arnt Myrstad

Abstract In discussions about Euclidean and non-Euclidean geometries, to take one central example, the notion of intuition plays a decisive role. According to Immanuel Kant, no geometry is possible without being based in a priori or pure intuition of space. The arrival of non-Euclidean geometries has generally been understood to falsify this claim, in particular according to Henri Poincaré's introduction of conventionalism in the philosophy of mathematics. However, even on the surface, qualifications are introduced into the discussion, such as "the intuition of representative space" and "rational intuition", indicating that the term "intuition" covers up serious ambiguities. Furthermore, the models of so-called non-Euclidean geometries, allegedly based in conventionally chosen sets of formal axioms, are directly or indirectly the very same objects claimed to be given to intuition in Euclidean space. The paper explores the nuances in the use of the term "intuition" in Kant and Poincaré, thereby discovering deep paralogisms in the discussion about geometry and hence resolving some of the problems pestering modern epistemology, by focusing on the border zones of conceptions covered by the same or similar terms.

The Unnoticed Complexity in the Conceptions of Intuition

A clear empiricist trend, and, in particular, an almost unanimous consensus that there are no synthetic a priori truths or principles to be had that can claim validity come what may, dominated the twentieth-century philosophy of science and general epistemology. Hence, even though there are still influences from his theoretical philosophy around, Kant's main project has for the most part been thought to be defunct. This trend started already in the late nineteenth century, and, ironically enough, in the group of philosophers called neo-Kantians. They linked the ongoing

J. A. Myrstad (✉)
Nord University, Bodø, Norway
e-mail: johan.a.myrstad@nord.no

J. S. Methi et al. (eds.), *Borderology: Cross-disciplinary Insights from the Border Zone*, Springer Geography,
https://doi.org/10.1007/978-3-319-99392-8_8

development in the sciences, and in particular in mathematics, to a need to let go of Kant's notions of space and time as pure intuitions. Subsequently the revolutions in physics, linked to the development of relativity theory and quantum physics, as well as the advances in mathematical logic, were seen by most as even more definitely and firmly ending any hope of finding and establishing synthetic a priori principles of knowledge.

However, at the end of the twentieth century and the first decades of the next, there has been renewed interest and belief in the necessity of constitutive principles that can make up the framework for developing, in particular, scientific knowledge. David J. Stump has recently summed up the status of the discussion, so far, in a book where he usefully revisits the attempts in the last century at explaining and legitimating the actual occurrence of such constituting principles or elements in all the "exact" sciences, i.e. logic, mathematics and physics: *Conceptual Change and the Philosophy of Science. Alternative Interpretations of the* A priori, Routledge 2015. The basic alternatives available according to Stump are substantive synthetic a priori principles, somehow in accordance with Kant's proposal, or conventional principles, somehow established by human fiat or, finally, originally empirically discovered rules subsequently turned into constitutive principles.

Stump's own preferred alternative is the last one, i.e. that constitutive principles are at core empirically founded and discovered, and that we may eventually treat them functionally as necessary for the establishment and development of a particular branch of science. However, Stump ends up admitting that none of the candidates seriously proposed as basic constitutive principles, such as the fundamental laws of logic, the axioms and postulates in the diverse disciplines of mathematics available, or the basic principles of mechanics, can plausibly be reconstructed according to this model. Furthermore, he also has to admit that one has not shown any mathematical truths to depend on such empirical knowledge in a way that make them fallible and actually falsified with the rapid expansion and correction of our empirical knowledge. Finally, the basic principle of classical physics, i.e. the principle of inertia, is still with us, e.g. as fundamental to the establishment of relativity theory and quantum physics. So, some of what Kant claimed to be basic synthetic a priori constitutive principles have stayed constitutive and unchallenged throughout the radical revolutions in the "exact" sciences.

However, since Stump agrees with the wisdom dominating last century's theoretical philosophy that synthetic a priori principles or elements in Kant's sense are impossible, he actually argues apagogically for his alternative by arguing against what he considers the only *prima facie* viable alternative, i.e. conventionalism. His general reason for considering the Kantian alternative impossible is a threefold one: (a) human fallibility, (b) historical change through so-called scientific revolutions á la Kuhn, and (c) that the discoveries of the principles or elements have taken place at particular historical times, and that we hence did not know these truths before those events. Now, this general reason is particularly disingenuous as launched against Kant, since Kant elaborated the events of the different scientific revolutions in the foreword to the second edition of the *Critique of Pure Reason*. He furthermore exposed and explained the proofs of the most basic constitutive principles (for empirical knowledge) in the part of that book called the *Analytic of Principles*, and laid bare the source of the endemic philosophical fallacies in the transcendental illusion exposed in the part called the *Dialectic of Pure Reason*. Hence, whatever force Stump's reason has, it should be confronted and argued thoroughly against the background of Kant's treatment of the phenomena it is based in.

Now, Kant claimed the even more basic principles concerning the notions of space and time as pure intuitions as well as what is presented in the aforementioned *Analytic of Principles* in the *Critique of Pure Reason*, are constitutive, not of natural science as such or mathematical science as such, but of ordinary experience ("Erfahrung") and perception ("Wahrnehmung"). Hence, their discovery as principles is linked to Kant's expositions of them in that work, the results of strenuous work through more than thirty years, and based on the more than two thousand years of thinking accomplished by numerous philosophers. However, their reign as constitutive for experience as such did not depend on their discovery. On the contrary, it was the reflection on their always already presence in and for experience that eventually led to their discovery as such.

Anyway, since most of Stump's arguments in his book are involved in opposing the supposed only viable alternative to functional a priori principles developed from empirical ones, that is, some version of conventionalism, he is throughout engaged in a discussion of Henri Poincaré's philosophy of science, the alleged father of this approach to fundamental principles in science, and only more indirectly with Kant.

Stump subsequently discovers that Poincaré always held a much more complex position and that the details in his position developed rapidly over the years. Hence, Poincaré, more or less explicitly, held that there are substantial synthetic a priori, as well as empirically discovered and conventional elements in the constitutive frameworks of the sciences. Furthermore, his reasoning about what makes the basic synthetic a priori principles possible, necessary, true and unavoidable, is highly original and interesting. Anyway, one cannot consider Henri Poincaré as an outsider; he was central in the development of the most advanced mathematics as well as natural science at the time, and his contributions to philosophy of science are still very much relevant for present-day discussions, as Stump's exposition of these certainly verify. However, as Stump's exposition directly and indirectly demonstrates, certain discrepancies in, and some unfinished aspects of his views, drove the mentioned rapid development and changes of his views, a process that, due to his untimely early death, was left unfinished.

Unknown to most, Kant's complete system involves a similar complexity as Poincaré's, i.e. it contains elements of substantial synthetic a priori character, as well as empirical ones refined into functional a priori ones, and conventionally instigated structures. However, the system is established in much more detail and secured a coherent unity since every element is determined a certain place bordering on and supporting the others. This paper will subsequently explicate some of the details of Kant's system and their architectonic positioning with respect to each other.

The border zone where the conflicts and tensions between Poincaré's and Kant's systems play out is defined by their diverse notions of intuition. As already mentioned, a decisive influence on the main, standard and dominant philosophical epistemologies developed or refined in the twentieth century comes from the so-called neo-Kantian movement, mainly active in Germany at the end of the nineteenth century, and their opposition to the Kantian notions of space and time as pure intuitions. In the English- and French-speaking parts of Europe this resistance was exacerbated by the connotations linked to the English and French terms "intuition" that are not or only weakly associated with the German "Anschauung" or their Scandinavian counterparts as, e.g. the Norwegian "anskuelse". While the English and French terms have a wide range of uses and even different meanings, none of these fully covers what Kant, in particular, had used the term "Anschauung" to express.

There are at least five different senses associated with the terms "intuition"/ "Anschauung" to be found in the work of Stump, Poincaré and Kant; one of them only found in Poincaré though Kant covers most of the sense or the reality behind Poincaré's use of the term with a different term, i.e. "die produktive Synthesis der Einbildungskraft". Then there is one sense that we only find fully expressed and exposed in Kant, though one may read Stump as attempting to cover it with the term "representative space" in the case of space as a pure intuition ("reine Anschauung"). Kant also operates with a whole range of different technical terms such as "formale Anschauung", "empirische Anschauung", "Raum als Form der Sinnlichkeit", "Raum als Form der Erscheinungen", "Raum als Form der Anschauungen", whose senses and uses cannot be found in the other two, and which are usually mistreated in the scholarly discussions, especially in English, about Kant's philosophy.

Stump apparently differentiates between rational, conceptual and propositional intuitions on the one hand and something one can best interpret, if imprecisely, as, an immediate, irrational hunch. As the notion of hunch indicates, such an intuition is supposed to carry no convincing truth-force, and has a sensible role to play only at the beginning of an investigation, not as the instance that closes it. Stump also thinks that rational, conceptual or propositional intuitions for the most part are chimeras, since they share the characteristics of hunches that they are immediate, unarticulated opinions-makers, without a link to any convincing grounds for believing the propositions, concepts or reasoning, with which they are associated.

Poincaré for the most part shares Stump's negative opinion of rational intuitions, in particular those versions of it, which he (falsely) thinks cover Kant's conception of "reine Anschauungen" as a base for Euclidean Geometry. However, he identifies a specific kind of intuition, which he claims is the foundation of all mathematics and which embodies a particular source of truth, not reducible either to that of logical argumentation, or sensible perception, and which one can neither identify with a correspondence or a coherence criterion of truth. This kind of intuition generates synthetic a priori truths, according to Poincaré.

On the other hand, Poincaré is at best unclear as to whether a similar form of intuition delivers synthetic a priori truths in geometry. However, as Frege made clear, there should be no conflict here of the evidence characteristic of arithmetic with the fact that truths of geometry depend on constructions in the (presentative) space of our perceptual intuitions. Only by completely separating geometry from the space of our perceptions can we secure that (spatial) intuition is not involved in "geometry". We simultaneously empty the field "geometry" of all sense and use as geometry, i.e. as a mathematical teaching about objects of (presentative) space. Poincaré never seems prone to accept such consequences, though Stump may tend to draw them on his behalf, as indicated above. An answer to the question why this is not something Poincaré himself would embrace, may be forthcoming from his explanation of the role of intuition in arithmetic: (Stump 2015, 50)

"On the other hand, Poincaré thinks intuition has a role to play in mathematics that cannot be eliminated. While he holds that intuition is unhelpful and even dangerous in geometry, Poincaré thinks that we know a crucial part of arithmetic by intuition—the principle of mathematical induction:

> Why then does this judgment force itself upon us with an irresistible evidence? It is because it is only the affirmation of the power of the mind, which knows itself, capable of conceiving the indefinite repetition of the same act when once the act is possible. The mind has a direct intuition of this power, and experience can only give occasion for using it and thereby becoming conscious of it. [...]

Poincaré seems as traditional here as Russell and Frege. Poincaré's attitude toward intuition in arithmetic presents a striking contrast to his view that intuition of the geometric primitives is completely empty. He thinks that arithmetic is synthetic a priori knowledge and that mathematical induction is the central principle in arithmetic. The issue I want to focus on for a moment is how he came to view geometry and arithmetic so differently."

Stump is here evidently on the search of a diagnosis of Poincaré's alleged failure to realize that he should have followed the tides of the time, and become a consistent and fulltime anti-intuitionist and formalist. On the contrary, below I will try to bring Poincaré's reflections on the subject's faculties and self-awareness in the case of mathematical induction in line with his reflections on the origin of the principles of mechanics, and in particular the principle of inertia, in the subject's self-awareness as a body, and hence as a solid object. Or rather, the other way around, bring Poincaré's thoughts about the human subject's perceptions of his or her body to comply with his thought about the subject's generation of the notion of whole numbers. The human subject knows about straight lines through her knowledge and apprehension of her own acts of drawing a straight line, i.e. a line in a specific and determinate direction. This act and our immediate awareness of it is as intuitive and certain as our apprehension of the succession of the whole numbers as governed by the principle of mathematical induction. The intuitions *of* space and time are thereby unfocused; they rather constitute the forms or the horizons for the actualization of the acts. They constitute the basic assumptions or conditions of the acts, tacitly accepted when acting.

The concept of natural number contains a notion of abstract succession, as in numeration. When applied to quanta in space and time it must relate determinately to the homogeneous successive expansion of a so-called extensive magnitude such as length or area, or similarly to an intensive magnitude such as the degree of the strength of a physical force or the intensity of a light source. According to Poincaré, the intuition associated with enumeration or number immediately validates the principle of mathematical induction. The source of our conviction of the necessary truths of arithmetic is our own consciousness when carrying out the enumeration.

According to Kant, and rather plausibly, the very same source of truth is at work when we draw a straight line in thought, or comprehend the uniform rectilinear motion of something. To demonstrate to ourselves through our own power of imagination the inertial motion or the inevitability of the truth of the inertia principle, i.e. that a material something that is not influenced by any moving forces will remain in rest or rectilinear uniform motion, has the very same characteristic self-reflective nature as the former examples. It also shares with the former the destiny that we will never be able to find an empirical realization of a perfectly straight line, or purely rectilinear uniform motion or strictly inertial behaviour. Stump therefore agrees that for this reason alone we cannot easily explain these principles and truths as the results of empirical principles converted into functionally a priori ones.

The Architectonic Complexity of Constitutive Elements in the Exact Sciences

Hence, we have these, in accordance with Poincaré's exposition of the principle of mathematical induction, self-evident insights. Since they were enacted and reflected into the propositional form of postulates they have been unchallenged and fundamental to all subsequent knowledge in mathematics and natural science, and in particular to the development of special relativity, in the version of Einstein's 1905 paper. In the first chapter of that paper, he establishes inertial systems in thought by erecting coordinate axes aligned with each other in the three directions pointing out space as three-dimensional, and in uniform, rectilinear motion with respect to each other. Thus, Einstein bases the further development of the theory in the very same fundamental insights that grounded the original conceptions of straight line, motion and inertia. The foundation of the theory itself therefore falsifies the claim that the theory of relativity has shown synthetic a priori truths and the structure of space founding the original conception of a straight line to be illusions.

The main question that Kant posed in the *Critique of Pure Reason* is hence still viable: How are synthetic a priori judgments possible? With respect to the drawing of a straight line (in thought) Kant linked his answer closely to his exposition of space as pure intuition or, rather, as "reine Anschauung", since it is very difficult to reformulate Kant's answer within the ordinary connotations that surround the English word "intuition", as already mentioned and to some degree explained above. Space as "Anschauung" is something within which we are immediately embedding, enacting and embodying ourselves, with whatever state or degree of articulation or consciousness. Hence, it is not something that we can correctly characterize as mental, as a representation within our mind. Rather it is the immediately given and unavoidable condition for the presentation of objects to us, including our own body, including all its parts and their spatial organization and reorganization as in movement. As such, we present space immediately to ourselves as the holistic unity of directions, i.e. the directions in which we direct our attentions to the objects around us and in which we enact our tasks, such as the drawing of a straight line in thought. Equally immediately and directly connected to the form aspect just mentioned, we give space to ourselves as the holistic unity of perspectives the objects can be approached within, and the viewpoints from which we may reach out to them, whether in perception or action. Only thus are the conceptions of the things we meet on in our everyday world comprehensible to us, as demonstrated with the Gestalt-switch figures that occupied and confounded the late Wittgenstein. When seen as a rabbit we immediately see it in accordance with our conception of a rabbit; i.e. that it has fur not feathers that it will leap to one side when startled that we will be able to discern its particular manner of movement from whatever side we see it and so on and on. The deep and seemingly endlessly detailed way in which we then organize our "picture" of the rabbit is through and through structured in accordance with space as such a holistic unity of directions, perspectives, viewpoints and aspects. Furthermore, none of these perspectives, viewpoints, and aspects will make any (determinate) sense without the as immediately, though necessarily unfocused, way in which we present our own body's placement and orientation, as something we embody, in the very same space as a holistic unity.

Since we distally focus, through our senses and actions, on objects and events presented to us in space, the proximally enabling embodiment of ourselves in viewpoints and perspectives is necessarily only preconsciously present to us. However, the evidence and thus Poincarean intuition enabling our knowledge of straight lines as lines drawn in one specific direction is based in the lines and directions that constitute viewpoints and perspectives that we enact in our perceptions, imaginings, and actions, as in the drawing of a line in thought. We can then validate these basic synthetic a priori truths only in strict first-person mode, i.e. as embedding, enacting and embodying selves, and any attempt to evade the responsibility for doing so will invalidate our opinions. We hence also base the definitions, postulates and demonstrations linked to these basic truths in the selfsame enacting and reflecting and articulating of these kinds of syntheses, a phenomenon that will be discussed more thoroughly in Chapter "Potential, Problems and Challenges of Joint International Master Programmes: Case-Study of the Joint Norwegian-Russian Master Degree Programme in Borderology" of this book.

The saying that we may see with our bodies, which usually refers to the manner in which we integrate the modalities of the different senses and in particular those (secondary) qualities that are linked to our bodies such as hardness, softness etc., with those that are available to our eyes, is thus much more fundamentally true. Similarly, the saying that we see with concepts, since we only see in a determinate sense what can be conceived of as in principle given to all perspectives, and from all viewpoints, and with aspects corresponding to those viewpoints and perspectives, must be turned around into a more profound one. All concepts that may ever refer to something given to us in perception and experience, must necessarily articulate the synthesis and synthetic unity that holds the different aspects of what we see together as given to us in (presentative) space. This is the manner in which concepts that shall mean something for our experience must necessarily have their basic determinations and hence definitions. All other forms of definitions, be they explicit or implicit, must base in this more fundamental one.

Stump is, as everybody, somehow aware of space in the sense explicated above, however, without clear consciousness of its irresolvable subjective-objective character. Hence, he only points vaguely to it under the denomination "re-presentative space". The manner in which he, as usual everywhere and anytime in such discussions, approaches spatial objects, is then also predominantly as given before us, with the determination and even existence of the "before us" left unmentioned and even covered up. When, as we all do, Stump is focusing on the spatial properties of geometrical objects such as spheres and hyperbolic surfaces, he, as we all do, immediately identify their characteristic spatial properties as we give them to ourselves in thought, or rather in space as "reine Anschauung". As soon as he, as Poincaré and most philosophers in the past have done, start asking critical questions about the definition of the concept of straight line, the evidence for the proof of the parallel postulate or its equivalents, or the demonstration of the principle of inertia, he gets himself into trouble. The only available alternatives for determining the concepts or terms seem to be either explicit definition, presupposing the meanings of the concepts and terms used to make the definition, or implicit definition through the assumption of a set of axioms with undefined non-logical terms. The only available alternative for proving theorems seems to be logical deduction from other theorems, finally based in assumed axioms and postulates. The only available alternatives for the acceptance of axioms and postulates seem to be either empirical verification or just some sort of hunch.

He, as we all regularly will do, has somehow forgotten to self-reflect upon the constitutive subjective-objective absolute unity of space as "reine Anschauung", resulting in the cover-ups, driven by diverse forms of paralogisms. Reminding oneself of the necessity to self-reflect on what we thus have forgotten is how Galilei understood Plato's *anamnesis*, an understanding that resulted in the establishment of the principle of inertia.

Now, Galilei's discovery and validation of the principle of inertia did not happen through empirical measurements and experimentation. On the contrary, as documented in the *Dialogue concerning the two main systems of the world* (1632), he had no experiment in the strict sense of the word based in measurements and a Cartesian system of coordinates to lean on. Because the measurement instruments, e.g. for time, such as pendulum clocks, that could measure time precisely enough and hence motion, had to be invented and demonstrated, partly based on physical laws themselves proved on the basis of the principle itself, such as the law of the pendulum. Instead, Galilei discovered, validated and defended the principle through a complex and involved conversation between three personifications of himself, in the dialogue called respectively Sagredo, Salviati and Simplicio, comprising a dialogical demonstration mostly based in simple everyday experiences and series of thought experiments. Later Kant developed the kernel of this demonstration into a comprehensive presentation in a short paper *Von den neuen Begriffe von Ruhe und*

Bewegung (1758), subsequently integrated into a systematic form in *Metaphyische Anfangsgründe der Naturwissenschaft* (1786). Anyway, we cannot find inertial motion anywhere in our world, neither in principle nor in reality, only more or less purely inertial motion and for the most part combinations of inertial and non-inertial, since influenced by physical powers, motions. Nevertheless, Einstein put the principle to use in his exposition of the special theory of relativity. Another component in this presentation is the erection of Cartesian coordinate systems for each inertial framework, at first only as thought or rather as exposed and construed in space as "reine Anschauung" through straight lines put together transversally on each other; straight meaning lines drawn (in thought) in one specific direction. So far, Einstein had called upon no (empirical) measurements. However, the choice of the Cartesian system of coordination, presupposing a conception of length as homogeneously extended along any direction, is already a conventional choice, since Einstein could have chosen to determine coordinates in a cylindrical or spherical coordinate system, just to mention a few of the more natural alternatives. When he had chosen the Cartesian coordinate system for each inertial framework, there remained for him to choose material measuring instruments for both time and distance measurements. Einstein assumed that he might accomplish this with the sort of instruments already available, or actually with any plausible alternatives that he knew from already gained empirical experience and experiments have physical natures and properties that make them well suited to be measuring instruments, of distances and time-periods, respectively.

So already this far into the exposition Einstein gave of the special theory of relativity, we have explicated three levels of constitutive principles or elements and, surprisingly enough, as successively based on each other in a systematic fashion. The conception of a straight line seems most basic, then follows the notion of uniform rectilinear motion, then the notion of inertial systems; furthermore the Cartesian system of coordinates, and then the choice of measuring instruments purportedly from physical experience shown to measure lengths and times correctly. Hence, operational definitions of lengths and times based in measurements made with the chosen measuring instruments, have as a constitutive presupposition; the definition of straight lines based in "reine Anschauung". The conventional choice of a Cartesian coordinate system is likewise based in the conceptions of straight lines and right angles. Furthermore, the conception and presentation to thought of right angles is itself based in space as the holistic absolute unity of directions, perspectives etc., i.e. in "reine Anschauung".

The next step in Einstein's exposition concerns the coordination of measurements made at different places in every inertial frame work. We are still within the realm of what we may call ordinary experience since Einstein thinks we can secure the coordination of the measurements of distances at different points in each framework, assuming that transporting a measuring rod from place to place slowly should not disturb the rod permanently. Similarly, we may distribute clocks of the same make to all places where we need time measurements.

At this point, however, we must take into account some extraordinary empirical results that in particular concern the coordination of time measurements. One has carried out many experiments that verify that the speed of light in vacuum is constant in all directions in space. One originally took this to indicate a medium of light propagation, wherein the speed is consistently the same in all directions. Hence, one suggested that carrying out more finely tuned experiments, among them, one famously called the Michelson-Morley experiment, one would be able to identify the inertial system that is at rest with respect to this light medium, the aether. However, the experiments failed to identify this system. Hence, one could claim that one can consider any inertial system as the rest system with respect to the aether, as measurements of light are concerned. Hence, on this empirically discovered fact, one could base a system for coordination of the clocks through light signals, ordering that all clocks in an inertial framework be synchronized according to the operational rule that the speed with which the signal was assumed to go from one point to another should be counted as constant, i.e. as c. Now, this is a determination that at least with respect to some aspects of it, Einstein constituted by a conventional choice. This is evident when we advance to the next step, which concerns the coordination, not only of clock measurements made within each inertial framework, but the coordination across inertial frameworks, where Einstein choose what is called the Lorenz transformations. This choice is not without alternatives, and this is seriously important for the philosophical discussion of special relativity, in particular as this has enfolded in the last 20 or so years.

Finally, one can calculate the predicted measurements of time readings and distance readings. These results depend then on all the constitutive elements described above. The results are numbers, for instance presumably indicating distances along three spatial coordinate axes, as well as clock readings measured in accordance with the ticking of the clocks and the synchronization processes and transformations chosen. It is essential for the system to work that we can agree on every time and place as identical for every inertial framework and system independently of these coordinates, which differ among the systems. Only thus can we coordinate the values of these variables given in the different systems. We can agree upon this as a common ground for all the inertial frameworks and systems just because these systems, with reference to any place and time, can identify coincidence in space and time as "reine Anschauungen", i.e. what Stump calls representative space and time. We will then have different measures for all these factors in the different inertial systems and frameworks, given by four real numbers, three for the measurements of the three spatial dimensions and one for time. We further correlate these measurements through predictions of what clocks and rods will show in the different systems at the same place and time for all systems. Hence, we once again presuppose (maybe without clearly understanding that, why and how) the basic holistic space and time as a unity of spatial and temporal å perspectives, directions, aspects and viewpoints, to make sense of what we predict we will find.

We can then treat the predicted results of measurements with purely arithmetical methods, i.e. ignoring how they in a complex manner represent the many levels of constitutive elements evoked. These elements comprise aspects of time and space as holistic unities of directions, i.e. as "reine Anschauungen", as we certainly realize in the form of the conception of motions between the inertial systems. They further comprise the dynamical physical properties of the rods and clocks used as measuring instruments, the synchronization procedures and the Lorentz transformations, e.g. in the case of the standard Einsteinian approach. Treated solely as arithmetical we can explore the measurement results with the help of different mathematical techniques as embedded in strictly arithmetical "spaces", such as what is referred to as R^3 or R^4, and we can demonstrate different properties of these arithmetical structures. These properties will naturally have parallels in our presentative space and time, since we have arrived at these predicted measurement results from the ground up by working with our imagination in presentative space and time by using, e.g. Cartesian coordinates.

Claiming that these properties somehow directly demonstrate that we need to discard our notions of presentative space and time, if one in any way thinks that this is at all possible, is seriously disingenuous for multiple reasons, though this has nonetheless been the dominant standard view for more than a century. I have exposed some of those reasons for not agreeing to the standard view above. More directly concerned with the results of the imagined measuring process, I will here only point out that the measurement results that are coordinated with the Lorentz transformations are gathered from different inertial systems. Each one of these inertial systems is using a synchronization procedure that assumes that light has constant one-way speed (in all directions) with respect to itself, and, hence, not with respect to any one of the other inertial systems in relative motion to the one chosen.

The underlying physical phenomena thus make it possible to consider each system as if it is at rest with respect to the unknown system, within which light moves with constant speed in all directions, something demonstrated with the experiment of Michelson-Morley and many others. In all other cases, we would only compare directly measurement results made in one coherent measurement system. However, we may translate results made in one system, as distances measured from one vantage point with those measured from another, by relating the vantage points to each other with respect to distance and direction and then translate the results measured from one into those measured from the other. This is exactly what the Lorentz transformation does.

Thus, one has demonstrated mathematically that we can choose one inertial system as the "rest system" consistently as the base for a synchronization procedure that rules not only clocks at rest in this system but also those clocks that move with the others. This gives a consistent and universal determination of all clock readings and rod measurements, indicating that matter in motion with respect to the rest system will shorten in the direction of motion and clocks in motion with respect to the rest system will slow down, all in accordance with the famous Lorentz factor. The details are available for those interested (cf. Guerra and Abreu 2008; Abreu and Guerra 2015; Andersen 2017).

The discussion about the interpretation of relativity will certainly continue. However, as argued here, it is decisive for the sound outcome of this debate that we observe, distinguish and reflect upon the different layers of constitutive principles, elements or aspects at play and how these interact and presuppose each other systematically. Only thus can we hope to get on to a more enlightened stage. By the way, Henri Poincaré, who had contributed much to the development of the special theory of relativity and recommended Einstein for his first academic position as professor in physics in Prague, late in life declared that he had been convinced that the aether theory of Hendrik Lorentz is the correct physical interpretation of the theory. This interpretation physically validates the aforementioned synchronization procedure in a way that in the results and formalism makes explicit the dependence on the original presentative space and time, the basic conception of motion, and the inertia principle.

The Paralogistic Confusions Created by Mixing Conceptions from Different Mathematical Disciplines

There are mathematical disciplines that explore different aspects construed in presentative space and time. Such as analytic geometry, which explores the nature of spatial figures in accordance with the algebraic and arithmetical structures they expose when treated with the help of, e.g. Cartesian coordinates. We have developed calculus that we use to describe more precisely the curvatures of lines and, e.g. the areas and volumes of spatial figures. Then we have algebraic topology, which with algebraic means can characterize the spatial form of, e.g. a sphere or a donut. Most of these disciplines we can generalize into arithmetical and algebraic forms with more than three "dimensions".

However, none of these various mathematical disciplines competes with each other for truth, as physical theories might. They are all validated more or less in the same manner, and they all seem immune to being falsified through empirical experimentation. Of course, some of these mathematical disciplines may deliver tools to the physical ones, and one may at different times and for different reasons consider one or the other more or less fruitfully applicable. This situation confounds Stump, since it goes against his declared attempt to reconstruct all constitutive principles and elements of knowledge as functional a priori ones, generated from empirically established knowledge.

The framework for Stump's conviction is the debate on the state of the mathematical discipline of geometry at the end of the nineteenth and the beginning of the twentieth century. This debate arose over the subject of so-called Euclidean and non-Euclidean geometries, a subject that one at the time scrutinized with the sharpened tools of modern mathematical logic, developed by Gottlob Frege and Bertrand Russell. As already mentioned, the background for the debate was the dismissal of the Kantian notion of intuition, by some, and the attempt at defending it somehow by others. Interestingly enough, Frege and Russell, who were central in developing the logic that others, such as Hilbert and Poincaré, used to argue more forcefully against the dependence of mathematics and in particular geometry on intuition, were themselves defenders of intuition as a source of (mathematical) truth.

In the next section, I will go into more detail on Stump's argumentation on this point, since most of the other aspects of the discussion about constitutive principles and elements in the exact sciences, as shown above, depend on this one. However, I will first present the general historical background for the discussion.

Geometry was the first exact science that one developed into a system in the form of a structure consisting of definitions, axioms, postulates and demonstrations of theorems. In the form delivered to posterity by Euclid's *Elements*, one had a model for the exactness and systematic character, which other sciences should strive to develop. However, there was at least one flaw with the system as Euclid had composed it in the *Elements*: the fifth postulate, the so-called parallel axiom, did not seem to have the same simplicity and evidence as the other ones. Over the centuries one carried out a multitude of attempts either to reduce the fifth postulate to the other four, usually by trying to derive contradictions from alternatives to the fifth postulate, alternatives claiming that through a point outside a straight line one may draw none or infinitely many straight lines parallel with the first one. On the other hand, one tried to replace the fifth postulate with others that appeared simpler and more convincingly evident. One initially developed none of these attempts on the assumption that the fifth postulate is false, and though one could deduce some very strange-looking theorems from the axiom systems containing one of the two mentioned alternatives to the fifth postulate, one never derived a contradiction. Anyway, many theorems in the *Elements*, one cannot derive strictly logically from the axioms, postulates and definitions. One needs demonstrations of a different kind, accomplished through presenting to oneself the geometrical objects, construing further figures and deriving more insights into their interrelations from what one presents to oneself in presentative space, as is exemplarily the case in the standard proof for the theorem of Pythagoras about the sum of the angles in a flat triangle.

In the *Discipline Chapter* of the *Methodenlehre* in the *Critique of Pure Reason*, Immanuel Kant gave his understanding of a mathematical system, comprising definitions, postulates, axioms and mathematical demonstrations. He claimed and demonstrated that we secure all these components of a mathematical system, as the *Elements*, through immediately presenting to ourselves the mathematical objects and properties in pure intuition. The second part of the nineteenth century saw the development of analytic geometry, projective geometry and other more abstract mathematical disciplines, as well as that of second-order predicate logic, developed by Frege, in connection with his attempt to derive arithmetic purely from logic. This led many scientists and philosophers to the aforementioned abolishment of (pure) intuition as a legitimate source of scientific knowledge.

With the use of the stronger logical tools available through second-order predicate logic, Hilbert could develop in a convincing manner a reconstruction of the corpus of theorems presented in the *Elements* from an extended set of definitions, axioms and postulates, wherein all proofs are logically valid. This success boosted two philosophies of mathematics, formalism and logicism that either claimed to reduce all mathematical truths to logical ones (logicism) or at least reduce or replace all mathematical demonstrations with logically valid deductions (formalism). The latter alternative, formalism, seemed the most prudent and plausible one, in particularly since Frege's attempt to reduce arithmetic to logic had met with some apparently unresolvable problems.

However, one still had to account for the definitions, the axioms and postulates without having to base these in (pure) intuition. One started out with the assumption that one could define, that is, determine the senses of, the so-called non-logical terms, i.e. those pertaining to the specific subject studied, implicitly and indirectly through the roles they play in the axioms and postulates. Furthermore, one claimed that the axiom system as such is not actually a set of truths but an abstract scheme for what one may find modelled by structures empirically verified as existing in the world. One had already developed the axiom systems with the alternatives to the parallel postulate, i.e. either the assumption that there are no parallel lines through a point outside of a straight line, or the assumption that there are many, actually infinitely many, such lines. One had subsequently demonstrated these systems to have viable models, e.g. the surface of a sphere or a so-called hyperbolic surface, respectively (in the two-dimensional cases). To stay with the more straightforward example, in the case of the sphere one then had to understand the (non-logical) notion of a straight line to point to grand circles on the sphere surface.

These models are certainly not empirical ones. The precision and exactness of a sphere and a grand circle on a sphere is not anymore available through implicit definitions and empirical examples than the original notion of straight line, and it does not demand any less the basic understanding of directions that pervade all our perceptions and imaginations, constituting them into determinate ones. Furthermore, one can hardly claim that the axiom systems indirectly determine their specific models and hence define the non-logical terms, since the so-called non-Euclidean axiom systems, i.e. those with the mentioned incompatible alternatives to the parallel postulate, have several different models that one can present to oneself in (pure) intuition. Actually, the models for the non-Euclidean axiom systems and their geometrical properties seem as dependent on (pure) intuition of (presentative) space as Euclidean geometry ever was.

Finally, with the advent of Gödel's theorems in the 1930s, one could demonstrate that any axiom system that somehow includes the knowledge of the natural numbers, cannot ever be completed in such a manner that all viable theorems are derived from the axioms by purely logical means (cf. Cellucci 2013). Hence, one cannot reconstruct mathematics with respect to definitions, axioms, postulates or proofs, as a legitimate and viable science in accordance with the tenets of formalism. All the same, the discussion of the foundations of mathematics during the last century does not seem seriously influenced by this fact, a situation I blame on the debunking of (pure) intuition still widely accepted. Below I present a more detailed discussion of the early parts of Stump's book where he takes the standard view on intuition and the foundation of mathematics for granted. My ambition is to demonstrate how he then implicitly and without noticing it, builds on (pure) intuition of presentative space, all the same.

Resolution of Some of the Enigmas Presented in Stump's Book

Stump's descriptions and characterizations of the models of non-Euclidean geometries are, as usual in these cases, carried out with the help of drawings of the spatial figures and lines, etc. allegedly demonstrated to fulfil the axioms of non-Euclidean axiom schemes, i.e. when the non-logical terms are re-interpreted in accordance with their new meanings. Stump, as his predecessors in this game, all the way back to Poincaré and others, presupposes that we determine and communicate these meanings with the help of these figures, and the words pointing out the relevant features of these. This is seriously disingenuous since Stump and most others involved in these matters think that one could not accomplish and communicate the determination of the original Euclidean meanings of the non-logical terms of the axioms in these ways. Actually, this was allegedly some of the motivation and the aim of the formalistic approach to geometry and mathematics in general. I.e. since one doubted that we could give the terms (precise and determined) meanings through our intuitive apprehension of spatial features, eventually pointed out with directing "fingers", one sought to define them implicitly through the logical structure of the set of axioms.

Originally, as Stump makes clear, one did not think there was a foundation crisis in geometry, along the lines sketched by the introducers and motivators of formalism. One naturally did not understand the notion of non-Euclidean axiom systems as an affront to and as contradictory or otherwise incompatible alternatives to the Euclidean ones. Actually, nothing should have been farther from their sound minds, since they used the very same spatial intuition to ground the determination of the non-Euclidean interpretations of the non-logical terms as they used to determine the Euclidean ones. I will point out that this is still the case with the introduction of models for non-Euclidean axiom systems by Beltrami, Klein and others. To shove this into your face I have italicized a few of all the appeals to spatial intuition Stump makes in his description of Beltrami's, and Klein's models: (Stump 2015: 13)

> Following the work of Gauss who developed a way to studying the geometry of surfaces, hyperbolic geometry can be considered a geometry on a saddle-shaped surface, that is, one that *curves* in two different *directions*. Klein provided a different model of hyperbolic geometry, in which the straight lines of hyperbolic geometry are considered as *open-ended line segments within a disk*. Taking these segments as lines, *we can easily see* that it is possible to *draw* more than one parallel (remember lines that do not *intersect*) to a giving [sic!] line through a point off of the line and that our strange theorem that given two lines that are *perpendicular* it is always possible to draw a line that is parallel to both *is obviously true*.

What should also be obvious is that since we can produce two quite different models in our spatial intuition—which supports our Euclidean geometry—for the non-Euclidean hyperbolic axiom system, that the logical strictures put on the non-logical terms in the latter system are not sufficient to give those terms a determinate meaning within a single model. Hence, there are also no reasons to think that the Euclidean axiom system is sufficient to determine the meanings of the non-logical terms. In fact, as demonstrated in the citation from Stump's book, what we need and what is sufficient to determine the meanings of the non-logical terms used in all kinds of geometry is the spatial intuition within which all our ordinary perceptions, experiences and imaginations take place. Furthermore, these take *place*, as we ourselves in our actions and explorations within the world, more or less independently of how far and wide our conceptual resources have developed and been made explicit in language.

As already noted, the interest in non-Euclidean geometries, or, rather, in alternative axiom systems to the Euclidean one, grew out of a bewilderment with respect to the fifth postulate, i.e. the parallel postulate, plausibly so since it had a less simple form than and hence seemed to point to a different sort of aspect of space than the others: (Stump 2015: 21)

> To keep things simple, I discuss only plane geometry (two-dimensional) [JAM: Spherical geometry and hyperbolic geometry as demonstrated respectively on the surface of a sphere and a saddle-shaped surface are also two-dimensional] and consider only one alternative to Euclid's geometry. The focus of long-standing discussion in geometry is parallel lines, that is, lines in the same plane [JAM: Hence, restricting the discussion to planes, not surfaces in general, already presupposes the determination of the surface as defined through straight lines in two different directions] that do not intersect [JAM: Lots of curvy and awkward lines can be drawn in a plane that do not intersect; what we need once more are lines drawn in determinate directions]. Euclid's version, the parallel postulate, (or axiom) says that if the line forms an acute angle to the perpendicular [JAM: Determination of the perpendicular also depends essentially on the determination of directions, as well as any angle] of a given line, it will eventually intersect the given line.

One may think that the lacunas in Stump's description, indicated by my block parentheses, are just inaccuracies. However, these lacunas are essential to give the impression that the case, with respect to what parallel lines are, is an open question until one has chosen one of the equally useful and valid available axiom systems. As my parentheses demonstrate, we have no way to make sense of what is expressed here without already presupposing our precise and evident understanding of what a straight line is, that is, of directions in space. Without immediately integrating into our perceptions, experiences and imaginations, our apprehension of directions, nothing empirical could ever make sense to us, and we would never have developed concepts of spatial objects and space and geometry, in the first place.

That is why we should be interested in alternative formulations of the axioms, and in particular the parallel axiom. Not because we could ever get rid of any of them by logical reduction to the others. On the contrary. Developing alternative axioms that together with the others would be equivalent to the parallel postulate gives us further insights into the specific aspect of our spatial intuition attaching to and somehow given expression in the parallel postulate: (Stump 2015: 22)

> For example, the assumption that there are similar triangles (triangles with the same angles but with sides of different length) is equivalent to Euclid's parallel postulate.

We need to be a little more precise: Space contains the possibility to present figures of the same geometrical shape, anywhere in space, and of all sizes. Thus, we express that space is infinite and homogeneous throughout. In this way, we have given a better and more adequate expression to the dependence of all spatial geometry on directions in (presentative) space. Stump, however, simply concentrates on the possibility of accomplishing a logical reduction of one postulate or axiom in terms of the others: (ibid.)

> Since the postulate and the replacement are equivalent or interchangeable, it is clear that if mathematicians were trying to eliminate the assumption of the postulate, they were not getting anywhere. Some explicit statement which is equivalent to the parallel postulate was necessary to proceed with the task of proving the theorems of geometry.

Again an imprecise expression by Stump (and he is following in the footsteps of many others) points us in the wrong direction. The parallel postulate or any of the equivalent replacements are necessary to infer validly all the theorems presented in Euclid's *Elements*. However, we have demonstrated many geometrical truths that we cannot derive in this manner. For instance, those treated in Apollonius' *On Conics*, projective geometry or what have you. More to the point, Kant demonstrated the necessary existence of incongruent counterparts in his ("Von dem Unterschied", 1768), completely outside of and independently of the Euclidean axiom system, solely from the evidence of the basic conceptions of the system of directions in space. These characterize (presentative) space as infinite and with a qualitative unity not evident in the Euclidean system as such.

Now, what does Stump think the demonstrated existence of Euclidean models of non-Euclidean axiom systems shows us? First, of course, that, as far as the logical structure is evident in the relations between the non-logical terms, these systems are consistent. However, Stump (again in the footsteps of many others) thinks that they show that our original conceptions of straight lines are vague and hence open to various interpretations: (Stump 2015: 25)

> That is, we can see that it is possible to have more than one parallel line because the models show us that there are surfaces or analytical constructs where it is true that there is more than one parallel. Therefore, the assumption that a line forming an acute angle to a perpendicular never intersects the given line does not lead to a contradiction, since there are objects in Euclidean geometry that exhibit these properties. Of course, taking these models to represent hyperbolic geometry assumes that it is possible to assign new meanings to the term straight lines in hyperbolic geometry. The debate over the formalism of geometry at the turn of the century is a debate over why we now consider it legitimate to redefine the terms of a formal system in this manner; that is why our preconceived idea of what straight lines is no longer was taken to rule out the possibility of there being more than one parallel line. The new conception of geometry allows for a change in what had been considered a priori knowledge.

Thus, just by using the very same a priori pure intuition that originally gave rise to the insights organized in Euclid's axiom system we can produce a model for the non-Euclidean axiom system in the very same Euclidean constructions of surfaces and lines, if we only change the meanings of some of the non-logical terms. Subsequently, someone (Stump among them) declared these a priori intuitive demonstrations to be the enthronization of a priori knowledge, in particular the specific synthetic a priori knowledge of Euclidean geometry, established within the very same a priori pure intuition as the models. Furthermore, they carried out this alleged revolution through what looks suspiciously like a hoax, i.e. just by changing the names of things. Anyway, the revolution seems to be a thoroughly reconstructed one, first concocted in the minds of some historians: (Stump 2015: 28)

> A revolution did occur in philosophy when formalism in mathematics took hold but interestingly many of the major actors in that movement do not see a connection to geometry. There is no revolution in geometry until stock is taken as to what happened, and this is true even in mathematics proper. For example, the development of spherical geometry was not considered revolutionary at all, because it was not seen to conflict with Euclid. In fact, spherical geometry conflicts every bit as much as hyperbolic geometry [...]. It was only when historians started drawing philosophical consequences from the development of geometry in the nineteenth century that the revolution which had already taken place could be seen.

What Stump states here is an expression of the internally inconsistent manner in which one considers "non-Euclidean" geometries. On the one hand, one sees them as contrary or even contradictory alternatives to Euclidean geometry. The reason one gives is that they obey formal logical axiom systems having at least one axiom that appears as a formal contrary to one of those in the Euclidean axiom system, e.g. the parallel axiom. However, on the other hand one originally demonstrated that these alternative axiom systems are consistent since one can model them in Euclidean (i.e. presentative) space, either on a sphere, a hyperbolic surface or a Klein disk. Naturally, then, no one at first thought that spherical or hyperbolic geometries contradicted Euclidean geometry (Stump 2015, 27–28). On the contrary, as Stump documents, the widespread opinion in mathematics as well as philosophy until after the turn of the century, was that there was no crisis in the field of geometry (Stump 2015, 28).

> [W]hat is important here is that there is no evidence that Riemann, Beltrami, or Klein were responding to a crisis in the foundation of geometry. Quite to the contrary, geometers seemed to have accepted the consistency of hyperbolic geometry and the impossibility of proving the parallel postulate even before the work of Riemann, Beltrami, Klein, and Hilbert had put the non-Euclidean geometries on a sound logical footing.

Stump mentions a very significant fact about the situation in philosophy of logic and mathematics around the end of the nineteenth century that we should scrutinize and ponder thoroughly. Those engaged most deeply and creatively in the development of formal logic (in particular second-order predicate logic), i.e. Frege, Russell and Wittgenstein (early and late), never thought that geometry could or should be reduced to the study of formal logical axiom systems, or that these systems gave an adequate treatment of the content of geometry. Since Frege and Russell, in particular, had worked hard on formalizing arithmetic, they certainly had acquired a deep understanding of the problems and limitations involved with such formalization. Hence, we should consider them not only as experts, allowed their saying in the matter, but also as those having the highest competence and skill with respect to formal logic and formalization. We need to explain why one has ignored their arguments and judgments in the aftermath of the alleged revolution concerning the foundations of mathematics, as we will attempt below.

As discussed above, the development of formal logic, and in particular the development of second-order predicate logic, gave enhanced instruments to formalize mathematics in general and geometry in particular, and to test axiom systems for consistency: (Stump 2015, 32)

> The consistency of hyperbolic geometry could be proved with new rigor, and all of mathematics could be seen as one large axiomatic system. This, new framework was the product of a new interdisciplinary coalition between philosophers and mathematicians, and the process of developing this interdisciplinary coalition is an interesting area to study.

The problem was that showing the axiom systems being consistent in this abstract sense did not really unify mathematics on the decisive levels. Since the abstract treatment left the essential sense and truth of the mathematics untouched, it rather contributed to a false feeling of security and unity. When Gödel demonstrated that formalism could not cover the truths of mathematics, the failure of formalism became even more evident. However, this did not much affect the opinions of the majority of those who worked on philosophical questions concerning mathematics. They were mainly still mesmerized, and are so still today, by the false story of the development caused by an alleged crisis in geometry (Stump 2015, 33).

> Both of these motivations require the development of non-Euclidean geometries to be described in a form that will be useful, and it appears that describing the developments as a revolution that occurred in response to a crisis was just what was needed to show the limits of rational intuition that classical mathematicians and philosophers had used, and to convince students that they had better learn how to express themselves in symbolic terms and to formulate rigorous proofs.

Within a similar atmosphere, Poincaré strived to develop his more complex and, I will argue, more advanced and adequate views on mathematics. However, Stump reads this “mixed account” more as a vice than a virtue. (Stump at times hence becomes overly demagogic): (Stump 2015, 40)

> Fundamental constitutive elements of science can change, leading to profound revolutionary changes in science. What Poincaré shows us is how change can occur at a very deep level that cannot be equated with mere empirical change, since his conventions are isolated from empirical testing. Of course, for Poincaré an even deeper bedrock remains a priori knowledge in the traditional sense in the case of arithmetic (in particular mathematical induction). Therefore, Poincaré has a mixed account, with some traditional a priori elements and some conventions, which can be taken as constitutive elements of scientific theories.

The mixed account of Poincaré has however to do with more than two levels of constitutive elements. On the one hand, we have those elements that according to Poincaré constitute the various scientific disciplines as such. The various Euclidean and “non-Euclidean” geometries are at first, as demonstrated through the models of spherical and hyperbolic geometry, such as Beltrami’s and Klein’s, based in the very same presentative space within which we organize our outer perceptions and what our perceptions are about. Furthermore, we have the different disciplines that construe various aspects and more or less abstract structures that can be related to pure objects in that space, such as continuous group theory, topology, and projective geometry. None of these competes with Euclidean geometry in the sense that accepting them makes the former false. Rather they complement each other and study different aspects and structures that we can construe in the presentative space within which we originally construed Euclidean geometry.

That Poincaré sometimes thinks that the only synthetic a priori element in mathematics is the whole number, we may explain by the standard paralogistic identification of metric geometry with the arithmetical theories of n-tuples, e.g. the Cartesian three-tuples with Euclidean three-dimensional space, we discussed above. Though we certainly may use calculations in such three-spaces or n-spaces to reveal properties of the original presentative space, we must establish this correspondence by reference to the straight lines, etc. that constitute the framework for the coordinates, etc. of points, lines, etc. in the original presentative space.

We have given citations from Poincaré where he justifies his belief in a truthful intuition as the base for arithmetic in the constitution of the faculties of the enumerating subject. There are only two steps from this to Kant's exposition of the pure form of time as the essential source of arithmetic knowledge, i.e. the revelation of the various ostensive and symbolic constructions involved and the specific basic transcendental schema "Zahl". Poincare cuts himself off from discovering a similar source in the subject for geometry, and eventually for the basic laws of mechanics, by, on the one hand, clinging to the at the time developing scheme of formalization in geometry, and, on the other, focusing on a description of the subject's physiology from the outside. He could have used his insight into the source of arithmetic as inspiration for a deeper understanding of geometry, had he seen clearly that non-abstract formal and abstract structures have their foundations in the non-abstract and purely intuitional. The purely intuitional constitutes the base for what we have secondarily structured through conceptualization, as in the cases of the various Euclidean and "non-Euclidean" geometries, group theory, algebraic and arithmetic topology, projective geometry and what have you. Subsequently Poincaré may have been able to reflect on how we constitute our indwelling in the outer world, and how we organize our sensations, faculties and perceptions into complementary fields of proximal and distal awareness, only through presentative space and time. Thus, he would have had to revise his description of the basis of our knowledge of original geometry, kinematics, and dynamics.

Finally, Poincaré would have had to look closer into the role of the inertia principle and its relation to presentative space and time: (ibid., 41)

> His thoroughgoing relational theory of space and his view that space exhibits only topological properties leads him to the view [that] the metric geometry of space cannot be determined empirically. This argument against empirical determination of metric is certainly the most important defense of Poincaré's conventionalism, and I discuss it later in the chapter, but Poincaré's argument also includes an a priori argument against Kant's theory of geometry, and this argument should be understood in the context of a philosophical theory of meaning and a formal conception of geometry that was discussed in Chapter 2.

That the context for Poincaré's argument against Kant is a philosophical discussion about the meaning of concepts and terms in the more specific context of logical formalization of geometrical axiom systems, as Stump here claims, is highly significant. The language philosophy of the twentieth century based in formalization has in parts muddled the ground for such a discussion, or, as we should rather hope for, dialogue. Kant's in many ways decisive and significant work concerning definitions, axioms and demonstrations in mathematics, already developed with considerable precision and comprehensiveness in the pre-critical work "Deutlichkeit" and then further refined and completed in the *Methodenlehre* of the *Critique*, has been generally ignored. By reintroducing some of Kant's insights into the debate, we may be able to solve the problems still pestering this discussion.

Meanwhile, Stump's ambivalence concerning Poincaré's two-fanged approach to conventions, and, furthermore, Poincaré's insistence on a kernel of synthetic a priori knowledge that cannot change during scientific revolutions, threatens to undermine Stump's own thoroughgoing empiricist and functional conception of the constitutive elements in our knowledge. The problem for Stump is ironically enough that Poincaré's conventionality of metric geometry is a real thing: (Stump 2015, 42)

> Missing the distinction between the conventionality of principles and the conventionality of metric geometry has led to many errors of interpretation, so it is important to elaborate the difference. The conventionality of principles is most analogous with the idea that some statements of physical theory function as though they were a priori or are constitutive, although Poincaré does not use this language.

Poincaré's conventionalism concerning metric geometry, on the other hand, means that one can develop the different metric geometries completely a priori, and that only the decision or determination concerning which metric geometry to use in empirical science depends on a choice. Furthermore, this is a choice, according to Poincaré, which one may always make so that Euclidean metric geometry is the one applied. Hence, neither pure geometry, whether as formalized according to Hilbert or as arithmetical according to R^3 or R^4, nor applied geometry can ever falsify Euclidean metric geometry, according to this form of conventionalism. A similar protection from falsification also applies to the basic laws of mechanics, according to Poincaré: (Stump 2015, 43)

> If these postulates possess a generality and a certainty which are lacking in the experimental truths from where they are drawn, this is because they reduce in the last analysis to a mere convention which we have the right to make, because we are certain beforehand that no experiment will come to contradict it […].

Hence, Stump's problems with making Poincaré a support for his own approach are getting stronger. There are no sign of arguing for the possibility of change here, as from fallibilism, for instance. On the contrary, according to Poincaré, no experiment will ever be able to contradict our choice of Euclidean metric geometry.

The true kernel of Poincaré's conventionalism concerning metric geometry is however the fundament of all metrical determinations in the original metric character of presentative space. This we can make exemplarily evident in the form of mathematical equivalence of the two rival systems of measurements applied to special relativity, as explored for instance by Guerra and Abreu (cf. references above). Furthermore, and maybe even more basic, there is the directional geometry tacitly introduced with the inertia principle, and hence incorporated through the central use of inertial systems in special and general relativity as well as in quantum physics. This also explains why Poincaré can be so sure that it will always be conventionally possible to reconstruct the physics in accordance with Euclidean geometry. The explanation is that we may guarantee ourselves this from the basic presentative intuitions of the form of our perceptions, actions, imaginings and objects. In the form of the inertia principle, we lift these presuppositions of our ordinary everyday experience into all our subsequent physical theories, the Newtonian ones as well as special and general relativity and quantum theories.

In our search for a basis, in Poincaré's reflections, of a profound revision of his outlook as a whole, we hence finally come to a very important passage. In this passage, which I have revised to make consistent sense of it, we find the blockage against any attempt to replicate his defense of intuition in arithmetic in the case of geometry and the basic laws of mechanics: (A repaired citation from *Science and Hypothesis*, Stump 2015, 43–44)

> At first blush, the analogy is complete; the role of experiment seems the same. One will therefore be tempted to say: Either mechanics must be regarded as an experimental science, and then it must be the same for geometry; or else, on the contrary, geometry is a deductive science, and then one can say as much for mechanics.
>
> Such a conclusion would be illegitimate. The experiments which have led us to adopt as more convenient the fundamental conventions of [mechanics] focus on objects which have nothing in common with those geometry studies; they focus on the properties of solid bodies, on the rectilinear [!] propagation of light. They are experiments of mechanics, experiments of optics; they cannot in any way be regarded as experiments of geometry. [JAM: Though they evidently presuppose basic properties of Euclidean geometry, i.e. exactly those related to the determined conception of and intuition of uniform rectilinear motion.] And even the principal reason why [mechanics] seems convenient to us is that the different parts of our body, our eyes, our limbs have precisely the properties of solid bodies. [JAM: One would then suppose that we had an independent access to the external properties of our own bodies as solid bodies. However, no such thing is the case. On the contrary, we have access to the properties of our bodies in the very same way as we have access to the properties of other outer bodies, through the pure form of our sensations making our outer perceptions spatially (and temporally) constituted, as well as of what those perceptions are perceptions. Here are the roots of Poincaré's most basic failures.] On this account, our fundamental experiences are preeminently physiological experiments, which focus, not on space which is the object the geometer must study, but on his body, that is to say, on the instrument he must use for this study [i.e. the study of geometry].
>
> On the contrary, the fundamental conventions of mechanics, and the experiments which prove to us that they are convenient, bear on exactly the same objects or on analogous objects [as those that make up the instruments for the study of mechanics]. The conventional and general principles are the natural and direct generalization of the experimental and particular principles.

However, the properties of solid bodies must already be characteristically metric to make possible a measurement of these metric properties. The system of measurement cannot be the origin of the sense of the metric concepts, and in particular not our immediate awareness of these metric aspects of outer objects. A simplistic paralogistic mistake convinced Poincaré that we could not ascertain metric properties through intuition. Contributing to this mistake is the confusion of the different conceptions of intuition involved that we have discussed above. Further contributing to this confusion is here the paralogistic confusion of our immediate awareness of our own body (as well as the outer objects of our perception) with the objects of physiological studies. Hence, Poincare lost the opportunity to understand the source of first-person active knowledge of space as similar to our first-person knowledge of whole numbers.

Anyway, the discovery of the principles of mechanics obviously were not the results of direct and natural generalizations of particular and experimental principles. On the contrary, Galileo needed more than 150 pages of the *Dialogue* to reveal the inertia principle, and then first make it evident and then subsequently defend it against natural and empirical principles of observations and (alleged) experience.

Furthermore, the long-cited passage makes it clear that Poincaré understood the nature of observation and perception as analogous to the pattern of measurement. A scrutiny of certain passages in Kant's "Träume eines Geistessehers" (1765) would have shown Poincaré (and others) that this is not the case. We might then lead them along the line of argument from "Träume" through "Von dem ersten Grund des Unterschiedes der Gegenden im Raume" (1768) into the dissertation from 1770 and the metaphysical exposition of space in the *Critique*. This would also have cleared the ground for a valid exposition of how terms refer and concepts mean. Alas, this we have to leave for a different opportunity.

As we have already hinted at above, and will scrutinize more thoroughly below, Poincaré's thoughts about the human thinker and observer as self-constituting may be worked out into a more consistent position allied with Polanyi and ultimately with Kant. Poincaré's conception of truth reflects this: (Stump 2015, 45–46)

> Instead of correspondence, Poincaré emphasizes the unifying of theory and of what is common to all thinking beings. These views may make it easier for him to claim that principles are true, but is not this also the same for the truth of geometry? Since Poincaré thinks we have clear alternatives in the case of metric geometry; he may have more reason for withholding the label 'true' from it. It is nice for my argumentation [JAM: Presumably for the distinction drawn between conventionality of principles and conventionalism concerning metric geometry] that Poincaré makes a distinction between principles, which are true, and geometry, which is neither true nor false, but can he maintain this distinction? It seems to me that this is problematic, given that to the extent that principles are empirical laws, they are true, but to the extent that they are conventions, they are neither true nor false.

It seems plausible to account for Poincaré's view concerning the truth or falsity of different parts of mathematics relevant to geometry, according to whether there are alternatives or not. If group theory or topology presented to Poincaré clear alternatives in the same way that non-Euclidean "geometries" do, the decision to choose from these alternatives would seem to be conventional. Actually, it only makes clear sense to call principles conventional when you really have alternatives from which to choose. It seems then that for Poincaré, the availability grounded in the subject's capacities, of such alternatives for a conventional decision, in similar ways as the principle of mathematical induction, does not stem from functional elevation of empirical principles, but is thoroughly a priori. In the case of the two-dimensional models of Klein and Beltrami, one can demonstrate the alternatives in presentative (Euclidean) space alongside the models of Euclidean geometry. For the alternatives of three-dimensional metric "geometry", we have no such possibility.

Hence, we have only the arithmetical demonstrations in R^3 or R^4 to make clear that the alternatives are a priori viable. Although the properties demonstrated in group theory, projective geometry and topology are without any obvious alternatives, we may treat them comprehensively in abstract symbolic constructions (this is the case for group theory and topology). Subsequently, we may easily demonstrate and apply them within presentative space (cf. in particular the example Euler discovered concerning the bridges of Königsberg, leading him to the initial stage of the development of algebraic topology as a mathematical discipline). However, we cannot present to ourselves in presentative space the non-Euclidean three-dimensional metric alternatives, in similar ways. We here need the use of physical models involving us with measuring instruments and their physically affected properties. Since Poincaré falsely seems to assume that the application of metric determinations involve the use of measuring instruments, he could plausibly claim that we cannot present the *metric* geometry, and thus the geometry he distinguishes from topology, etc. in presentative space. Furthermore, since he, as most of the interlocutors in the debate around the turn of the nineteenth century, falsely identifies presentative space with visual representations, he would be inclined to claim that metric geometry is not available to the kind of intuition permeating our visual experiences.

Hence, Stump's attempt to explain the availability within geometry both of a priori determinations (variably throughout Poincaré's development identified as group theoretical, projective geometrical, topological, and finally formally logical, respectively), and conventional choices, barks up the wrong tree. Partly because he ignores the role of presentative space, and concentrates on the ontological and abstract distinction between substantial and relational theories of space. However, the subject of relational versus substantial theories of space is certainly a background for other relevant aspects of the discussion: (Stump 2015, 46)

> As an alternative, I propose that Poincaré thought that a fully relational theory of space is not only possible, but that in outline it is quite defensible. Thus, in *Science and Hypothesis* Poincaré argues that the concept of absolute space is both "repugnant to the mind" and refuted empirically [...]. As many authors have noted, a fully relational theory of space would allow for conventionality of metric, an argument that I develop in the following.

Since a relational theory of presentative space is a mirage, one may wonder on what grounds such a claim is justified. However, it probably seems plausible that when one attempts to comprehend the spatial properties and directional relations with a hypostatical conception of space, one is (rather) free to make that comprehension in any way one chooses. To Kant a conception of space as absolute in the manner presupposed in this debate, is as repugnant as the relational dissolution of our spatial form of intuition. When one understands Poincaré as denying any intuition "of" space and time, allegedly thereby contradicting Kant, we must interpret this as a denial of intuition of the space and time of geometry and physics, not of presentative space and time. Otherwise, we will have a hard time (!) making sense of Poincaré's utterances: (Stump 2015, 46)

> First, he argues against any a priori determination of the metric of space, which is one of the many ways that Poincaré directly opposes Kant. He argues that we have no intuition of time and no intuition of space […]; completely rejecting Kant's ground for arithmetic and geometry. The existence of consistent non-Euclidean geometries thus shows us that there is no a priori method of determining the metric of space. [JAM: Here the ambiguity concerning whether we are talking about the models of Klein and Beltrami construed in presentative space, or the R^3 or R^4 models of the abstract formal axiom systems in arithmetic (metric) three- or four-"space" is devastating.] Poincaré also has what looks like a formal view of geometry, which may seem incongruous, given his strong anti-formalist stand in arithmetic, but nevertheless we can find this in his writings.

One needs first to distinguish between presentative space (and time) and the "spaces" of what Poincaré calls "metric geometry", i.e. R^3 or R^4. Only the former is truly and intuitively given, and the form of our outer perceptions and imaginations. The latter we can only make sense of as "geometry" when we link the arithmetic determinations to truly geometrical ones through rules of correlations installed with the help, e.g. of a Cartesian set of coordinate axes and rules for associating each point in presentative space with a coordinate three-tuple (x, y, z). Then we have to distinguish both these fields of determinations from the formal, logically structured, but with the primitive non-logical terms un-interpreted, structures of axiom systems. We may apply these axiom structures to all kinds of interpreted formulas, concerning all kinds of various fields of knowledge or nonsense. We may use them to formalize and study the logical structure of talks of apples and pears, imaginations and traumas and so on and on. Since we already have some axiom systems, i.e. the Euclidean and non-Euclidean ones that one has discussed for centuries, we may also formalize and study these systems. For instance, we may search for complete formal deductions in valid logical forms of all theorems from the alleged axioms, and suitably extend and supply the axiom set with lots of others to accomplish this task. In essence, this is what Hilbert did with respect to the Euclidean axiom system and the theorems developed in Euclid's *Elements*.

Anyway, we may subsequently attempt to apply these so-called Euclidean and non-Euclidean axiom systems either on structures discovered and researched in presentative space or in R^3 or R^4 or any other arithmetic n-dimensional "space". Actually, this is what one already had accomplished for the case of two-dimensional models of non-Euclidean axiom systems, i.e. the systems of Bolyai and Lobachevskii, that is, the models in presentative Euclidean space on spheres, hyperbolic surfaces and disks. Hence, one may correctly call these two-dimensional geometries, spherical and hyperbolic, respectively. To avoid paralogistic ambiguity, however, one should have renounced from calling the axiom systems themselves hyperbolic and spherical. The problems become aggravated when one applies the names also to the three-dimensional models in R^3 or R^4 of these axiom systems. Then, the problems finally become practically and historically unsolvable with the introduction of the so-called physical models of these arithmetic models that one may define through correspondence rules with the measurements made with imagined physical measuring instruments, such as metal rods, imaginatively affected by imagined temperature gradients or gravitational forces. Alternatively, and less purely imaginatively, as in the case of relativity theory, as discussed above, introducing some empirical knowledge about measurements of the speed of light.

Poincaré implicitly and tacitly keeps track of the presentative space in which he construes his physical model of a sphere with temperature gradients and the measurements made with metal rods affected by temperature, resulting in giving him the freedom to choose "metric" or rather arithmetic alternative "geometries" for the measurement system. Hence, in practice, he could handle the distinctions between what is empirical and what is a priori and what is conventionally determinable. However, he did not clearly note how he accomplished this, and, furthermore, he never, or apparently, only very late, were able to see through the confusions mentioned above, and state the distinctions drawn succinctly. For Stump, and others, this is much more difficult. Stump, for one, as already mentioned, has at least noted an important distinction in Poincaré's thoughts about conventionality: (Stump 2015, 46 cont.)

> Second, he also argues against any empirical determination of the metric of space, a point in the argument where Poincaré is often thought to be using something like the Duhem Quine underdetermination thesis. However, this cannot be the correct interpretation of Poincaré's geometric conventionalism, because all empirical theories, and not just physical geometry, are underdetermined in the Duhemian sense.

Hence, Stump has also introduced the distinction between pure geometry and physical geometry, at least *prima facie* corresponding to the distinction we have drawn above between arithmetic or "metric" "geometry" and its application to physical models in presentative space. However, subsequently Stump misinterprets the "pure geometry" term as referring to the formal un-interpreted "non-Euclidean" axiom systems: (Stump 2015, 47)

> Concerning pure geometry, Poincaré holds the modernist sounding views that we have no pre-axiomatic understanding of geometric primitives, that rigor demands that we eliminate appeals to intuition in geometry, and that pure metric geometry is neither true nor false. By calling his views modernist I am referring to the fact that philosophers and historians have recognized that Poincaré expressed a view of geometry that was influential in overturning the traditional view of geometry as a synthetic a priori science of space.

The development of arithmetic went on for thousands of years, independently of the discovery of the Peano axioms (already to some degree discovered and developed by Johann Schultz, "Hofprediger und Professor der Mathematik" in Königsberg, and a close friend and collaborator of Kant). The development of Euclid's geometry, Appolonius' theory of conics, projective geometry, calculus etc., etc., went on without anybody bothering too much about a problem with giving sense to the primitive terms. The supplementation and correction of the set of axioms of Euclid by Hilbert, supposedly necessary to give the primitive un-interpreted terms an implicit sense, was a very late accomplishment in the history of geometry. This makes up an immense amount of acquisition and application of what we then and always have called geometrical knowledge (in physics and technology as well as in everyday fields such as architecture and shipbuilding, constructions of instruments, locomotives and railways, and so on and on). On this impressive background, it is hard to take Poincaré's alleged views, as expressed in Stump's words here, seriously.

However, we may reserve the use of the term "pure geometry" for what we have identified above as formal logical axiom systems with un-interpreted primitives that we eventually may model in presentative space as what we always and originally have called geometrical figures. Then it makes some sense to say that the meanings the primitive terms are given are those that we can speculate they have through the restrictions they must obey because of the formal axioms in which they are used to express whatever sense or meaning these axioms have. Alternatively, but not equivalently or complementary, we may define the primitives through their applications in the arithmetic "pure" "geometries" of R^3 or R^4 or whatever.

These arithmetic models of Bolyai's and Lobachevskii's axiom systems in three or higher dimensions cannot themselves be directly modelled in our ordinary presentative three-dimensional space, through which we observe and imagine spheres and hyperbolic surfaces, as the two-dimensional versions can. However, when first imagining the ordinary geometrical figure of a sphere, and supplying this with a physical environment consisting of metal rods and temperature gradients, we may find a physical model, consisting of the measurements made with these metal rods considered as measurement instruments, of either Bolyai or Lobachevskii three, or four or what have you, dimensional metric arithmetic "geometries". By judicially coordinating the naming of terms, etc. we can show that these arithmetic theories, when applied in this way to physical systems, obey the "non-Euclidean" axiom systems and thus have interpretations, which obey the strictures these axioms put on the primitive un-interpreted terms. As demonstrated above, no one can ever get away from presentative space (and time). However, muddling the waters with equivocations of concepts and terms that are rather innocent in the practical works of mathematicians and scientists, since the terms get their relevant diverse meanings when applied in their different contexts, becomes endemic in the hands and minds of philosophers. Anyway, it is then also the responsibility of the philosophers to clear up the many paralogisms involved and relieve us of the confusions.

Conclusion

There are obviously more for us to do concerning the diverse problematics that have been touched upon in this article. The essential teachings we should take and carry with us into this work, concern the necessity of searching for the basis of scientific knowledge in the preconditions of our ordinary enacted, embedded and embodied everyday experience. These preconditions we cannot easily discern and make explicit. On the contrary, since these are what we are when we engage in the world, they must be employed, but for that reason not focused on, as such. Hence, the work of reflecting and articulating, and refining and developing our practical knowledge, can only take place by working through long periods of confusion and failure. A central, yes, indispensable tool to accomplish the evolution out of the confusion is the confrontations between engaged and honest thinkers in the border zones of their understanding, as I have attempted to do with Henri Poincaré and Immanuel Kant in this article concerning the diversity and structure of constitutive principles or elements of the exact sciences. However, we can only release the powers these thinkers can give to the dialogue, if we ourselves remain honest, and open to the sources of insight hidden in our own practical knowledge.

References

de Abreu R, Guerra V (2015) Speakable and unspeakable in special relativity: time readings and clock rhythms. Elect J Theor Phys 12:183–204

Andersen F (2017) Experience and theory: a defense of the kantian a priori and Kepler's philosophy of science in light of modern space-time physics, PhD Thesis 2017: 70, Norwegian University of Life Sciences, Faculty of Social Sciences, School of Economics and Business

Cellucci (2013) Rethinking logic, logic in relation to mathematics, evolution and method. Springer

Guerra V, de Abreu R (2008) On the consistency between the assumption of a special system of reference and special relativity. Found Phys 36:1820–1845

Stump DJ (2015) Conceptual change and the philosophy of science. Alternative interpretations of the a priori. Routledge

Deconstruction of European Environmental Identity in the Mass Immigration Context Summary

Basia Nikiforova

Ecology must stop being associated with the image of a small Nature loving minority

Felix Guattari.

Abstract The article is concerned with the relationships between the natural environment and identity in the conditions of ecological crises and mass immigration. Environmental identity is part of the common concept that people will identify with what they care about. Using Susan Clayton methodology approach, we underline that environmental identity is part of getting to know part of our self, the key to recognizing the part of our self that is related to the environment is in viewing ourselves as part of the environment. An environmental identity can be similar to another collective identity (national or ethnic) giving us a sense of connection, possibility to be part of a larger whole, and with a recognition of similarity between others and ourselves. The author offers to look on the European environmental identity development through rationality, growing solidarity, and responsibility. An environmental rationality is not just a philosophical and theoretical enterprise but is rooted in social practices, recovered sense of the being, restores the link with life, and formulates new arguments for mobilizing collective actions. At the same time, nowadays, the ethic of otherness is not a dialectic of opposites that results in the exclusion and elimination of the opposite (radical) other. In new materialist framework, nature is not a passive social construct but "an agentic force that interacts with and changes the other elements, including the human" (Alaimo and Hekman). The influence of the environmental change is largely unrepresented in standard theories of migration. Climate change is expected to have some bearing on future migration. Every attempt to identify a climate change migrant or a climate refugee in the present is almost impossible. The reason for this is that migration is not caused by any single variable. The important question is how climate migrant is constructed in the media. The figure of the climate change migrant is regularly constructed as a hapless victim or as object of fear or threat. The article deals with definitions of environmental migrants, environmentally displaced people, or climate refugees as "people who must leave their homes

B. Nikiforova (✉)
Lithuanian Culture Research Institute, Vilnius, Lithuania
e-mail: nikiforovaphilosophy@gmail.com

J. S. Methi et al. (eds.), *Borderology: Cross-disciplinary Insights from the Border Zone*, Springer Geography,
https://doi.org/10.1007/978-3-319-99392-8_9

and communities because of the effects of climate change and global warming". Environmental identity will be analyzed in the context of Felix Gvattari and Arne Næss ecosophy.

Keywords Deconstruction · Environmental migrants · Environmental identity New materialism

Introduction

Ecological crisis is particularly revealing subject for humanitarian and social research, especially in the present era of growing globalization. Ecological crisis is a process, where past and future are permanently clashed. The past was never simply there to begin with, and the future is not what will unfold, but "past" and "future" are iteratively reconfigured and enfolded. From the view of new-materialists, the past is open to change, "it can be redeemed, productively reconfigured in an iterative unfolding of space, time, matter" (Dolphijn & Van der Tuin 2012). Ecological crisis does not inherently determinate relationship between past, present, and future. It is a possibility to repair the now situation and the ability to change the past. These kinds of diffraction patterns possibly make changes in ecological public consciousness.

Two major new materialism authors Karen Barad and Donna Haraway explains how diffraction is a method for reading and writing based upon the physical phenomena. Diffraction is a way of coping with epistemological problems of representation. To paraphrase Haraway, diffraction is an attempt to make differences while recording interactions, interference, and reinforcement. For Van der Tuin, "diffraction is meant to disrupt linear and fixed causalities, and to work toward more promising interference patterns. This kind of diffraction patterns is not simply reflections, it does not displace them elsewhere, and instead, it produces changes in public consciousness. "The diffraction can be a metaphor for another kind of critical consciousness" (Dolphijn & Van der Tuin 2012: 26).

The ongoing biggest European refugee crisis is an example of such kind of diffraction which in reality is not only a concentration of current wars and conflicts but also represents negative ecological consequences such as poisoning and lack of water, air, destruction of property, terrain, landscape, and sometimes climate changes.

From Fred Dretske view who was a theoretic of epistemology, an agent is a being with the capacity to act, and agency denotes the exercise or manifestation of this capacity. Term agency is used in a much narrower sense to denote the performance of intentional actions. The philosophy of action provides us with a standard conception and a standard theory of action. The former construes action in terms of intentionality, the latter explains the intentionality of action in terms of causation by the agent's mental states and events. In the book "Explaining Behavior: Reasons in a World of Causes" he analyzes reasons from which persons

—agents with reasons—move in the way they do and how ordinary explanations, explanations couched in terms of agent's reasons, explain". (Dretske 1988: 88) For Dretske, it was important how triggering these reasons in real life and how this hierarchy is built. He was sure that every kind of human action was resulted by sum of reasons, waiting, and beliefs.

The epistemological approach is very important for development of environmental identity from the reason that it investigates practices in specific cultural and historical circumstances.

Environment as Identity Bedrock

The definition of identity is based on three principal notions: identity is formed and supported through internal identification of oneself with one's own ethnic, cultural, or religious group; identity is formed and supported through external identification, distinguishing ourselves from others; and other does not exist outside of me. It is the only unique way of one's self-identification. A person starts the search for identity beginning from the search of other within himself or herself.

For instance, after "regaining their independence, the Baltic States have been constructing their political identities in terms of the East/West opposition. They have been creating narratives of belonging to the West, with the East as their threatening *other*" (Miniotaite 2003: 214). Some researchers (B. Anderson, E. Gellner, and E. Hobsbawm) consider that identity is something constructive, imaginary, made up. B. Anderson claims that ethnos is an artifact, which has been generated by cultural and political leaders, but the case of Eastern and Central Europe strongly confirms the reality of ethnic communities (Anderson 1983). According to Zygmunt Bauman, modernity constructed the concept of identity and postmodernity was occupied with its semantic destruction. From his point of view, identity is still a problem, but it is not the same problem that was urgent during the entire modern period. The problem of identity in the modern epoch was that of how to construct identity and keep its integrity and stability. The problem of identity in the postmodern epoch is how to avoid inflexibility and to preserve the freedom of choice. National identity is formed as a symbiosis of national statehood and the ethno-cultural ground. The dominating criteria for distinguishing differences of national identities are linguistic and religious specificity of a concrete group. They render the deepest influence on maintenance of consciousness of the personal national identity, development and strengthening of its originality, and visible difference from other national and ethnic people and groups.

Our paper is concerned with the relationships between the natural environment and identity in the conditions of ecological crises and mass immigration. First of all, identity as such is a core psychological construct a way of describing an individual that locates him or her within a social and political context. Environmental identity as a part of common concept has different approaches such as environmental communication, ecological psychology, conservation psychology, and others. At the same time, each of them includes social and moral worldview, attitudes, and behaviors. If we decided that the identity is therefore constructed through interaction with others, we should also agree that ecology and environmental nature presents itself as other. From this view, we should agree that the key to uniting ourselves with nature is by recognizing ourselves as indissoluble part of the environment.

The environmental identity is possible to be divided through horizontally and vertically dimensions. First means that we have to deal with the sum of ideas, beliefs, imaginations, myths, and prejudices about natural environment. Second means that identities begin in our social environment where feeling of space, time, culture, and moral gives meaning to our constructed identity about nature as such.

Using Susan Clayton methodology approach, we underline that

- environmental identity as part of getting to know part of our self, the key to recognizing the part of our self as part of the environment.
- an environmental identity can be similar to another collective identity (national or ethnic) giving us a sense of connection, possibility to be part of a larger whole, and with a recognition of similarity between ourselves and others.
- the European environmental identity development through rationality, growing solidarity and responsibility.
- an environmental rationality is not just a philosophical and theoretical enterprise but is rooted in social practices, recovered sense of the being, restores the link with life as a whole, and formulates new arguments for mobilizing ecological collective actions.
- nowadays ethic of otherness is more tolerant, really it is not a dialectic of opposites that results in the exclusion and elimination of the (radical) others (genders, race, continents, and economies).
- contemporary ethic of otherness is more tolerant, really it is not results of the other' exclusion and elimination (genders, race, religion, continents, and economies).

Posthuman Identity According Rossi Braidotti

In the beginning of twenty-first century, the most well-known theorist who contributed to the rhetoric on posthumanity is Francis Fukuyama with his work "Our Posthuman Future" (2002). Fukuyama's analysis was about a future where human biotechnological enhancement starts to be the frame and definition of this future. Using negative arguments about posthumanism, he turns this problem more in the political than moral direction. For him, "human nature shapes and constrains the possible kinds of political regimes, so a technology powerful enough to reshape what we are will have possibly malign consequences for liberal democracy and the nature of politics itself" (Fukuyama 2002:7).

One of the most famous new materialism researchers Rossi Braidotti describes posthumanism as "the historical moment that marks the end of the opposition between humanism and anti-humanism and traces a different discursive framework, looking more affirmatively towards new alternatives" (Braidotti 2013: 37). From her view, posthumanist perspective goes further in exploring alternatives, without sinking into the rhetoric of the crisis of man. At the same time, it works towards elaborating alternative ways of conceptualizing the human subject. Man becomes the dominant subject as much as he includes his core characteristics and as much as he excludes others. Braidotti considers that the great emancipatory movements of postmodernity are driven by the resurgent others: the women's rights movement; the anti-racism and de-colonization movements; the antinuclear and pro-environment movements. These social and political movements are simultaneously the symptom of the crisis of the subject and also the expression of positive, proactive alternatives. Also, its existence are evidence of the crisis of the former humanist center position and dominant subject position ignoring environment and nature. "In the language of my nomadic theory, they express both the crisis of the majority and the patterns of becoming of the minorities" (Braidotti 2013: 37-38). This hopeful post-nationalist approach expresses the decline of Eurocentrism as a historical event and calls for a qualitative shift of perspective in our collective sense of identity.

Rossi Braidotti looks on posthuman identity as a way to bypass contradictions between a globalized and culturally diverse Europe on the one hand, and the narrow and xenophobic definitions of European identity on the other. She underlines the fact of Europe becoming nomad entails resistance against nationalism, xenophobia and racism, and bad habits of the old imperial Europe. Braidotti adds that it is the opposite of the aggressive universalism of the past, which is replaced by a situated perspective. Posthumanist project embraces a new political and ethical project, which reviving tolerance as a political and psychological tool of social justice and develops a new kind of post-nationalist European identity which partly includes characters of environmental identity. She concludes that "one of the most pointed paradoxes of our era is the tension between the urgency of finding new and alternative modes of political and ethical agency for our technologically mediated world and existed inertia of established mental habits on the other which is in reality general feature of our common identity" (Braidotti 2006: 36).

Ecology and New Materialism Framework

Ecotheory is a convergence of ecology, philosophy, anthropology, sociology, literature, feminism, sustainability studies, environmental justice (including indigenous and postcolonial studies), queer theory, and numerous adjacent fields that seeks to deepen understanding of the close relations of humans and nonhumans. In new materialism framework, nature is not a passive construct but "an agentic force that interacts with and changes the other elements, including the human" (Alaimo and Hekman 2008: 7).

Jane Bennett bases her political ecology of things upon an ontological narrative intended to restore the enchantment of the world in the agentic encounter between humans and nonhumans. She underlines the important element in the conceptualization of a material ecocriticism: the challenge, which the material turn poses to ecocriticism consists in the need to explore new narrative dimensions and new forms of narrative agency. If the concept of agency is no longer limited, it is almost impossible not to see material agencies as makers of stories. Bennett underlines that using the definition assemblage we have the possibility to methodologically unite such different objects as a human and nonhuman, nature and culture, body, and consciousness. An assemblage start to be a method for dialectical understanding world as a whole in their manifold and strong dependence. In interview "Agency, Nature and Emergent Property", Bennett remarks that an "assemblage" is grouping of an ontologically diverse range of different kinds of vital materialities" and every emergent process and event (blackout, hurricane, mass migration of human and nonhuman beings, and animal's epidemic) put to the common framework (Bennett 2009: 92). The assemblage as such gives the possibility to highlight some of the limitations in human-centered theories and to understand some practical implications which are very important for a public culture and actions and responsibility for the human—nonhuman separation.

Bennett introduces such definition as vibrant matter, vibrant materiality, vital materialism, or knotted world, and others with some tasks: "to open a horizontal prospect of reality or to reduce the ontological distance between the human and the nonhuman, to extend the horizon of values about the complex landscape in which the human action is situated" (Bennett 2010).

But what is exactly the meaning and importance of the new materialism in new conceptualizing environmental materiality and identity? Jane Bennett takes a step further for explanation of ecological new materialism framework. In her vision, life and nonlife, human and nonhuman, are only different forms through which matter emerges in its agentic capacity. It is like organic and nonorganic forms, are differential becoming in the entanglements of agentic matter. From her view, the vitality of matter rooted in the immanence of a scientific and ecological vision. Bennett's conclusion is not only ethical but it has strong political dimension: "In a world of vibrant matter, it is not enough to say that we are 'embodied'. We are, rather, an array of bodies. If more people marked this fact more of the time we continue to produce and consume in the same violently reckless ways?" (Bennett 2001: 112–113). Her book "Vibrant Matter" is a "Creed" for a post-environmentalist political project: "I believe that encounters with lively matter… highlight the common materiality of all that is, expose wider distribution of agency, and reshape the self and its interests" (Bennett 2001: 122).

From the view of Stacy Alaimo, environmental identity founded on such characteristics as:

- the crossing territories of environmental materiality and of social perception and construction,
- trans-corporeality radicalizes the view of trans-locality, and pictures the transit maps between the inside and the outside of every bodily form;
- trans-locality traces the flows and migrations of people, goods, and identities across places;

- trans-corporeality is a conceptual descriptor for the flows of substances and discourses across bodies;
- reconfiguring the boundaries between the self and the environment, as well as between all material substances in material contexts (Alaimo and Hekman 2008; Alaimo 2010).

All these characteristics produce a common image of environmental identity and show the boundaries reconfiguring between the self and the environment.

Felix Guattari: Triplex Discourses of Ecology

How does Felix Guattari help us rethink ecological crises and such its manifestations as the pollution of water and the air, the destruction of forests, the disturbance of climates, the disappearance of a multitude of living species, etc.? Felix Guattari offers a new conjunction of environmental ecology, social ecology, and mental ecology. This "new conjunction" gives start to the invention of new collective assemblages of enunciation concerning the couple, the family, the school, the neighborhood, etc. (The Guattari Reader 1996: 263). Guattari interactive interdependence triplex discourses of ecology include nature, society, and psyche. Here, Guattari turns to face toward the creative change and future new forms of solidarity. He underlines that our environmental crisis includes wider spectrum of political, social, moral, and gender dimensions:

> Our survival on this planet is not only environmental damage but: by a degeneration in the fabric of social solidarity and in the modes of psychical life, which must be reinvented. The refoundation of politics will have to pass through the aesthetic and analytical dimensions implied in the three ecologies—the environment, the socius and the psyche. We cannot conceive of solutions to environmental crises without a mutation of mentality, without promoting a new art of living in society. We cannot conceive of international discipline without solving the problem of hunger and hyperinflation in the Third World. We cannot conceive of a collective decomposition of the socius without a resingularisation of subjectivity, without a new way of conceiving political and economic democracies that respect cultural differences. We cannot improve the living conditions of the human species without to improve the feminine condition. Production for the sake of production—the obsession with the rate of growth, whether in the capitalist market leads to monstrous absurdities. The only acceptable is the production of a subjectivity that is auto-enriching its relation to the world in a continuous fashion (Guattari 1995: 20–21).

"Ecosophy" as Instrument for Environmental Communication

Norwegian philosopher Arne Naess described ecosophy as a research theory and discipline that is based on analytical thinking and examines nature and our relationship to it. An ecosophical concept being part of a personal worldview determines the interaction of a person with the nonhuman natural world. For him, an ecosophical concept which does not have the status of a scientific theory, but, first of all, governs the process of the formation of moral values related to the human interaction with the nonhuman natural world.

If we will try to compare Arne Naess and Felix Guattari ecological conceptions, we will find that the difference between key ideas lies in their understanding ecosophy as such. For Guattari, as we mentioned before, ecosophy can be of three types: environmental, social, and mental ecology. As a result of which we should protect not only our environment but our mentality and our psyche. The future of the world belongs to human possibility to change the types of subjectivity.

Naess' ecosophy concept is linked with the possibility to grow and develop personal worldview and a personal code of values. For him, to affirm the intrinsic value of every life form is ecosophy' general goal. Both of them confirmed the necessity of economic, political, and social reforms. But if Naess pays more attention to ecosophy' moral issues, Guattari in his turn—to social and political.

Fifty years ago, Arne Naess wrote that term argumentation patterns is not often applied to people (Philosophical Dialogues 1999: 444). He left recommendations for objective public debate which necessarily reminds today when we communicate and present refuges subjects. In the book "Communication and Argument", Naess recommended in the public debate to avoid tendentious statements of fact, irrelevance, ambiguity and quoting, and information which should never be untrue or incomplete. He underlined necessity to avoid tendentious tone of presentation (irony, sarcasm, pejoratives, exaggeration, and subtle (or open) threats. These are famous and simple rules violated in mass media during some last years.

Environmental Migrants in an Interconnected World

Climate change is expected to have some bearing on future migration. Any attempt clearly and distinctly identifies a climate change migrant or a climate refugee in the present is impossible. The reason for that is the following: migration is not caused by any single variable, important question is how climate migrant is constructed in the media, and the figure of the climate change migrant is regularly constructed as a hapless victim or as object of fear or threat. Next important question is that discourse on climate change and migration can be implicated in wider questions about race and religion. Postcolonial approach is very important in nowadays situation where animosity towards migrants is so prevalent. Structuralism paradigm of the

European identity means that an orientation to Europe derives fundamentally from association with other Europeans. Identity arises from interacting with others and coming to the realization that one has much more in common with them.

We deal with definitions "environmental migrants", "environmentally displaced people" or "climate refugees" as "people who must leave their homes and communities because of the effects of climate change and global warming". Statistics is critically dependent on the definition of "environmental refugees", a process which might well be seen as impossible given the multiple and overlapping causes of most migration streams. The influence of the environment and environmental change are largely unrepresented in standard theories of migration, whilst recent debates on climate change and migration focus almost entirely on displacement and perceive migration to be a problem. Norman Myers described environmental refugees as "persons who no longer gain a secure livelihood in their traditional homelands because of what are primarily environmental factors of unusual scope" (Myers and Kent 1995: 18). In his book "Ultimate Security: The Environmental Basis of Political Stability" which was called "provocative", Myers added that environmental conflicts will become the "principle threat to security and peace" in the years ahead (Myers 1993).

An initial difficulty in dealing with "environmental refugees", or "environmental migrants", is different typologies of them. Thirty years ago, Jon Martin Trolldalen, Gaim Kibreab, and others researchers (Trolldalen et al. 1992) distinguished between refugees from natural disasters; degradation of land resources; involuntary resettlement; industrial accidents; the aftermath of war; and climatic changes. Environmentalists found common cause even if the linkages between environmental change, conflict and refugees is only a hypothesis. The subject of environmental migrants necessary to examine in a wider context (borderology and conflict studies). Territory and borders are very important actors in this subject of research. It is possible to find connection between environmental and political refugees, because environmental degradation often directly linked with ecological consequences of wars and conflicts.

Deleuze and Guattari make a difference between such definition as nomad and migrant. They add that "the nomad distributes himself in a smooth space; he occupies, inhabits, holds that space; that is his territorial principle. It is therefore false to define the nomad by movement. It is therefore false to define the nomad by movement". From their view, "the migrant leaves behind a milieu that has become amorphous or hostile, the nomad is one who does not depart, does not want to depart, who clings to the smooth space left by the receding forest, where the steppe or the desert advances, and who invents nomadism as a response to this challenge" (Deleuze and Guattari 1987: 381). Perhaps, in this sense, environmental refugees are closer to the nomad then to the migrant.

Gaim Kibreab tried to give an answer on the question: why the term environmental refugee has been so seductive? From his view, "the answer lies in the agenda of policy-makers, who wish to further restrict asylum laws and procedures, at least in part to depoliticize the causes of displacement and enabling states to derogate their obligation to provide asylum" (Kibreab 1997: 21). Today, we observe this tendency as very clear.

Conclusions

We live in the time which radically changed the relationship between the natural environment and identity. Today, the ecological crises and mass immigration start to be a process on the same line. We should recognize that environmental identity as a part of our self, but at the same time is a part of environment. An environmental identity can be similarly important to another collective identity (national, ethnic, regional or religions), giving us a sense of interconnection, possibility to be part of a larger whole, and recognize the similarity between others and ourselves. We look on the European environmental identity and our possibility to develop social and economic life through rationality, growing solidarity, and responsibility. An environmental rationality, solidarity, and responsibility are not just the theoretical constructions but rooted in social practices and recovered a sense of the being and mobilize future generations for collective actions.

The posthuman theory develops a new kind of post-nationalist nomadic European identity that challenges bordered nation-bound identities. Environmental identity means that life and nonlife, human and nonhuman, are only different forms through which an agentic force interacts with and changes the other elements, including the human. Discourse on climate change and migration should be implicated in wider questions about territory, border, race, and religion. We underline that environmental migrants, or climate refugees are "people who must leave their homes and communities because of the effects of climate change and global warming". The environmental migrants' problem necessary to examine in wider social and political context.

References

Alaimo S (2010) Bodily natures: science, environment, and the material self. Indiana University Press, Bloomington

Alaimo S, Hekman S (2008) Introduction: emerging models of materiality in feminist theory. In: Alaimo S, Hekman S (eds) Material feminisms. Indiana University Press, Bloomington, pp 1–19

An Interview with Jane Bennett by Gulshan Khan (2009) Agency, nature and emergent properties. Contemp Polit Theory n.8/1:90–105

Bennett J (2001) The enchantment of modern life: attachments, crossings, and ethics. Princeton University Press, Princeton

Bennett J (2010) Vibrant matter: a political ecology of things. Duke University Press, Durham and London

Braidotti R (2006) Affirmation versus Vulnerability: on contemporary ethical debates'. Symp: Can J Cont Philos 10/1:235–254

Braidotti R (2013) The posthuman. Polity press, Cambridge
Deleuze G, Guattari F (1987) A thousand plateaus: capitalism and schizophrenia. University of Minnesota Press, Minneapolis and London
Dolphijn R, Van der Tuin I (eds) (2012) New materialism: interviews & cartographies. Open Humanities Press, an imprint of Michigan Publishing
Dretske F (1988) Explaining behavior: reasons in a world of causes. The MIT Press, Cambridge, MA
Fukuyama F (2002) Our Posthuman future: consequences of the biotechnology revolution. Profile Books, London
Guattari F (1995) Chaosmosis. An ethico-aesthetic paradigm. Indiana University Press, Bloomington and Indianapolis
Guattari F (1996) Remaking social practices. Gary G (ed) The Guattari reader. Blackwell, Cambridge
Kibreab G (1997) Environmental causes and impact of refugee movements: a critique of the current debate. Disasters 21(1):20–38
Myers N (1993) Ultimate security: the environmental basis of political stability. W. W. Norton & Company, New York
Myers N and Kent J (1995) Environmental exodus: an emergent crisis in the global arena. The Climate Institute, Washington
Miniotaite G (2003) Convergent geography and divergent identities: a decade of transformation in the Baltic States. Cambridge Rev Int Aff 16(2):209–222
Nina W, Andrew B (1999) Philosophical dialogues: Arne Næss and the progress of ecophilosophy. Rowman & Littlefield Publishers, Lanham, MD
Trolldalen JM, Birkeland N, Borgen J, Scott PT (1992) Environmental refugees: a discussion paper. World Foundation for Environment and Development and Norwegian Refugee Council, Oslo

Part III
Knowledge on the Frontiers; Conceptualization of Practice (Activity)

Man and Nature: Approaches to the Delimitation of Concepts

Viktor R. Tsylev

Abstract The very phrase «man and nature» implies a significant difference or even an opposition between these two forms of the living world existence. At first glance, the difference is obvious: man came out of nature, which is determined by the natural scientific laws, and built up his own world of culture, in which life is governed by social norms and laws. But we encounter some serious problems trying to draw a well-defined border between man and nature. These problems manifest themselves in two fields: in man (in opposing of biological and social bases of his behavior) and the world as a whole (in comparison of its two parts—natural and human (social, cultural)).

Keywords Nature and man · The biological and social in human being
Nature and society

The Problems of Differentiation of Biological and Social Bases of Human Behavior

First, let us consider the problems arising while we are trying to differentiate the natural and social in human behavior. The most common way here is making a boundary to separate the human characteristics similar to animals from those that are peculiar to man. In this case, the biological characteristics of human includes hereditary traits; instincts (of self-preservation, sexual, etc.); emotions; natural needs (in breathing, eating, sleeping, etc.); other physiological features similar to those of the mammals (functioning of internal organs, constant body temperature); as well as such abilities as adaptation to the environment, procreation, and so forth. That is, human biological characteristics include everything that is inherent for the living body as a representative of the mammals' species. The social characteristics include the ability to produce tools, articulate speech, language, social needs

V. R. Tsylev (✉)
Murmansk Arctic State University, Murmansk, Russia
e-mail: vrts@mail.ru

J. S. Methi et al. (eds.), *Borderology: Cross-disciplinary Insights from the Border Zone*, Springer Geography,
https://doi.org/10.1007/978-3-319-99392-8_10

(communication, friendship, love), spiritual needs (morality, religion, and art), the ability to think, to carry out conscious creative activity, and so on. In fact those are all of the features of human life that are caused by society and acquired by a person in the course of life. If a human does not undergo the socialization process, he does not develop the social features, as evidenced by the known cases of children sprouting among animals (as Mowgli).

With such a distinction of biological and social traits in a human being, we leave outside the scope those biological features of man that evolved during the anthropogenesis and thus are peculiar to humans—upright walking, developed hands, large, and advanced brain, the abilities to articulate speech and tool use. Those social features have already become a part of human nature as biological characteristics, and this fact allows natural philosophers to conclude that the social is essential for human nature and manifests itself as a naturally determined stage of human development.

If we associate the biological basis of human behavior with the physiology and body structure, and social basis—with the process of socialization, the point of this division is actually the presence of the biological substrate, and it is assumed that the social characteristics do not arise out of the biological basis directly and can evolve in humans only by means of socialization. And it is the social basis of behavior that is considered a specific human characteristic, while the physiological basis of behavior is attributable for many mammals.

This emphasis in the understanding of the social was set even by I. Kant. The specifics of a purely human behavior is seen by Kant in man's ability to determine his actions, being guided by reason, in contrast to the natural behavior, when a person is subject to the natural feelings or passions. Kant points to the qualitative difference of the sociocultural organization from the natural one and finds its basis in the reasonable and moral-spiritual essence of man. Authentic being is the destination and necessity, and not just an actual situation (Kant 1995). From this perspective, the social characteristics of human are not some immanent, initially inherent in man abilities, but the imperative that is yet to be realized by everyone.

K. Marx also tried to bridge the gap of social and biological in man. He thought that the natural biological form of human existence serves as the premise and the basis for his further development: «Man is directly a natural being. As a natural being and as a living natural being he is on the one hand endowed with natural powers, vital powers—he is an active natural being. These forces exist in him as tendencies and abilities—as instincts» (Marx 1971). The natural can be viewed as a form of human existence, getting its social content in the process of development. That is, the very nature of man contains the prerequisites for the social characteristics' development. This idea of Marx has been further developed in the Russian psychological theory of activity, in research of S. Rubinstein above all.

S. Rubinstein wrote that socially oriented behavior is performed by human mental activity, which is inextricably linked to physiological processes and morphological structures of the brain. In his reflex theory based on a large number of the experimental data, he shows that mental activity, as a higher nervous activity, has physiological and neurodynamic basis, reflected in a specific morphological structure of the brain (Rubinstein 1957). At the same time, morphological structure of the brain does not only determinate certain mental activity but is also dependent on this activity, being changed according to the performed function. Man assimilates socially evolved modes of action in the process of communication, training, and education by means of their stereotyping and turning into «generalized system of reflex connections fixed in the brain» (Rubinstein 1957), which actually leads to changes in the morphological structure of the brain. According to S. Rubinstein, the natural involves not only structural and morphological features of the brain but also the reflex activity of the brain. In turn, the conditional reflex activity of the brain is «formed in the process of individual development, being determined by conditions of practical and theoretical activities of man» (Rubinstein 1957). Thus, we can conclude that the development of new abilities in humans leads to the change of their own nature, since the attaching of these abilities takes place at the physiological (neurodynamical) level, being accompanied by the formation of the corresponding reflex activity of the brain and thereby being fixed in its morphostructure.

It is important to note that changes in human nature in this aspect are not due to its spontaneous development, but due to the assimilated modes of action. And internalized mode of action is determined not by social norms and rules, but by the updated human nature. For example, after a child has learned to hold a fork with confidence, it becomes easier for him to eat with a fork than with the hand, because this action is already being provided by his changed nature. S. Rubinstein concludes that the opposition of the natural and social (internalized) is wrong, because the very nature of man, i.e., his natural ability, developing in the process of ontogenesis, is formed as a result of education and his own activity (Rubinstein 1957). If we do not consider social behavior as having a biological basis, then social ways of actions after their assimilation cease to be social because they become biologically determined.

S. Rubinstein actually brings human assimilation of action modes to the development of his abilities that allow their implementation. In his development of abilities theory, he shows that these abilities have «organic, genetically fixed conditions for their development in the form of *inclinations*. People from birth are endowed with different inclinations, although the differences are not so great <... > Differences between people's inclinations are, above all, in the innate features of the neuro-cerebral apparatus—in anatomical, physiological, functional features» (Rubinstein 1998). According to Rubinstein, all humans have inclinations, formed as a result of the cultural-historical process. These inclinations are actually prerequisites for development of abilities, which are implementing the common ways of social behavior.

The presence of the inclinations to certain modes of action indicates that the ability to perform them is developed by the modern man much faster than it used to be hundreds or thousands of years ago. That is, human nature at the level of physiology has the potential of fulfilling certain social actions. But the abilities do not develop from inclinations by themselves; they have to be demanded by the environment: «Congenital characteristics of the organism are only very ambiguous inclinations, *specifying,* but *not determining* the mental qualities of man. On the basis of the same inclinations different qualities (abilities and personality traits) can be developed, depending on the course of his life and activity» (Rubinstein 1998).

In situations when a child (like Mowgli) grows among animals, there is no need to develop abilities of social actions implementation (they are not demanded by the environment), therefore these abilities do not develop. On the other hand, the presence of the inclinations (at the physiological level) to perform common social actions indicates that these actions are due to the presence of some physiological structures in the brain even before the development of abilities for their implementation. Perhaps, the same was the way of fixing in human nature of his biological features, evolved in anthropogenesis and already considered to be exclusively human: upright walking, developed hands, large, and advanced brain, the abilities to articulate speech and tool use.

Thus, it can be concluded that on the human level, the boundary between the social and biological are very conventional. Social ways of action acquire the physiological basis in the process of their assimilation, therefore, in fact, becoming biologically determined. Furthermore, the traditionally existing social forms of action have already been laid in human nature in the form of inclinations and are in fact partly naturally caused. At the same time, the implementation of these actions until their assimilation requires conscious control, so at the mastering stage, they should be considered to be more social than biological. It turns out that the pure social (nonbiological) in man includes those of human lifeways that have not yet been reflected in his physiology, being fulfilled under the control of consciousness. Since individuals acquire various action modes and abilities to a different extent, the boundary between the social and biological bases of behavior can be determined for each individual only. Thus, the problem of distinction of the natural and social in man is extremely ambiguous, and the boundary is movable, because, first, the same human actions and abilities can be considered as purely social or having a biological basis, and, second, this border is constantly changing and is determined individually, since assimilation of abilities and ways of action and their physiological fixation takes place in accordance with the individual life path.

Those conclusions have some far-reaching consequences. It turns out that human nature is being historically formed by the cultural and social conditions. Accordingly, the people living in different cultures are of different nature. For example, rationality has become a feature of the Western European man. This means that it has integrated into his nature (at the level of inclinations), and therefore is quickly assimilated by children, and for a man of this culture, it is easier and more natural to behave rationally than the other way. Then, it turns out that the instrumental, or rational, action (according to Weber's classification) for the Western European man has become a traditional action, because it has become a solid part of Western culture and actually expresses the change of human nature (Weber 1996). At the same time, migrants of other cultures, arriving in the Western Europe, have a different nature (at the level of inclinations) and can realize rational behavior only by means of the conscious control, so they can behave rationally for only some limited time periods, and it may take more than one generation to assimilate such behavior. Another behavior is habitual for them, the one that is caused by the cultural and historical traditions from which they came out, so, in fact, they require a different social and legal support in their lives.

On the other hand, a long-term residence in another culture causes changes of man's nature according to the conditions of life in this culture. A good example here is the history of the United States of America. At the time of the state formation, there was a gulf in the cultural and historical development of the white and black population, which determined a significant difference in the characteristics of their nature. Now, we can see the actual alignment of these groups and their representatives equally successfully participate in all spheres of national life. But it took more than 200 years to receive such an alignment because human nature is stable and changes just gradually from generation to generation with education and accumulation of new inclinations.

A significant difference in the man's nature can be formed in social and ethnic groups living in the different environments. For example, this difference can be seen between people living in tundra and the urban population. In the summer of 2016, I conducted a study in the tundra of the central part of the Kola Peninsula: I studied human interaction with nature, and among other things, I was watching the people living there: the reindeer herders and specialists related to reindeer husbandry (cross-country vehicle drivers and tractorists). Life in terms of the Kola tundra affects the character of the people: they are mainly modest to living conditions (can easily put up with the absence of shower, handy toilet, comfortable bed and bed linen, food variety, etc.), somewhat rude in communication, physically strong and hardy, steady for the cold, attentive to different natural events and signs of nature, and most importantly, accustomed in their activities to rely not on the regimented schedule, but mostly on the current environmental conditions. These qualities are required for living in the tundra quite comfortable. And there is another important feature of these people—most of them are capable to «feel» the tundra, to a greater or lesser extent, and in this respect, they can be seen as part of the tundra nature.

And, here, we come to the comparison of the human living in the tundra and the tundra nature. Let us use this example for consideration of the problem of distinction between «man» and «nature».

The Problems of Differentiation of the Nature's Life and Human Activity

P. Berger and T. Luckmann may seem to set a clear boundary of the social: social order is a human product or, more precisely, a continuous human production. Social order is not a part of the «natural order» and does not appear according to the «laws of nature»; it exists only as a product of human activity. It is a human product both in its genesis and in its present (Berger and Luckmann 1995). In this perspective all made by man is considered to be social. But the problems are beginning to arise when we try to define the degree of man's autonomy of the nature in his constructive activity. Can a created social order be regarded as a purely human product, independent from the influence of nature? To explain the essence of the problem, let us turn to the example of the interaction between man and the tundra nature.

At first, certain characteristics of the tundra nature should be noted. The severe nature of the tundra limits animals in respect of the livelihood resources, so the relationship of fauna in the tundra is particularly clear. In the years when the number of lemmings increases, the number of foxes and polar foxes increases too. And it is not the result of a causal link, it occurs simultaneously. The period between reindeer's rutting and calving in stable, it takes 223–227 days. At the same time, the arrival of spring in the tundra may vary considerably, and move up to 2–3 weeks. Despite this, according to the reindeer herders, the calving always starts at the time when the hills begin to thaw. That means that the rutting season, taking place in September and October, is clearly related to the time of spring arrival next year, when the calving starts. How is this correlation possible? It is difficult to assert the existence of a causal link. Let us give one more example: before the bad weather, the tundra animals hide themselves, as if being warned. There is an impression that all the living things in the tundra are linked, and they all get some signals for the fulfillment of certain actions.

Such natural phenomena are well explained from the perspective of the organic integrity philosophy of Russian religious philosophers of the Silver Age. This worldview sees the nature as consisting of two components: the visible (material) and the invisible (spiritual). The visible part is presented in a variety of physical forms, but this variety is a symbolic expression of the unified spiritual origin (Berdyaev 1994). The spiritual unity of the natural world is alike the unity of the human body (as the likeness of macrocosm and microcosm): in spite of all the variety of physiological processes and structural elements of the body, they are all interrelated for fulfillment the activities of the single «I» of a person. Similarly, the natural world acts as a single living organism. The sense of activities of the One lies far beyond the individual processes and its meaning is incomprehensible (Andreev 1992). For example, if the cloudberries crop is not good this year, it is a trouble for man and he blames the wind that tore off the blossoms. But from the perspective of tundra as a whole, cloudberry crop was not meant to be that year and that is why the wind tore off the blossoms. But if the sense of nature as a whole is incomprehensible, it is still possible to feel the processes occurring in nature, to follow them or somehow take them into account. The flora and fauna of the tundra feel these processes and fully obeys them, so everything in the tundra nature occurs smoothly and synchronously. And people living in the tundra are as well forced to submit to this process, in order to survive in the harsh conditions.

As a result, the orientation on the tundra nature (especially on the weather), on the current situation is a characteristic feature of the reindeer herders and others living in the tundra. Their planning is usually approximate and their plans can change quickly due to the changing weather conditions or the occurrence of any new unforeseen circumstances. In some cases, they make corrections to their plans in anticipation of changes in the weather or circumstances, not yet knowing exactly about them, but feeling the need to make these corrections, explaining their actions by some rational-constructed reasons. These situations serve as an indication of their inclusion in the tundra nature, where they act as a part of it. And in these cases, the drawing of the boundaries between man and nature becomes extremely challenging. On the one hand, reindeer herders are obviously people, and on the other hand, they act as a part of the tundra nature, joining its harmony and obeying its signals.

A similar incident happened to me while returning from the tundra after the study. I was with my family (wife and son) at the Tundra research station of Murmansk Arctic State University. Just over a month, I had been examining the social interaction of man and tundra nature. The case is in the fact that my wife and son came back home from the tundra by helicopter, which came for us on the day before the planned arrival, and I was left in the tundra and got back by cross-country vehicle; that was an unexpected way of departure for us and it happened with such a large number of concurring factors, that it is impossible to treat them as random.

We were to be taken along with a passing flight, and if there were some problems with the arrival of the helicopter on that day, then our flight would be delayed for a week. On the day of the planned arrival of the helicopter, a cross-country vehicle was also going from the tundra to Lovozero, so it was important to decide in advance what transport we would return by. Two days before the planned arrival of the helicopter, I tried to call the airport manager using a satellite phone to make sure everything was going as it had been planned, but after several unsuccessful attempts, my satellite phone unexpectedly ran out of money. On the other side of the lake, there was a cross-country vehicle, ready to leave that evening to the reindeer herders' base (situated in 4 km from our camp) and to return back the next morning. I asked the driver to make a call to the airport from the base and then give us information about the helicopter the next morning.

The next day was the day before our planned helicopter departure. We waited for the cross-country vehicle till 15.00 but in vain. That is why I decided to wait till 17.00 and then to go to reindeer herders' base to get the information about the helicopter, if the cross-country vehicle did not come back. At 16.25, all of us—me, my wife, and son—suddenly felt that I had to go. Remembering that mood, I could characterize it as a compound of readiness to leave and feeling the necessity of leaving at that moment. I packed up quickly and left at 16.30. Usually, it took me about 1 h and 20 min to get to the reindeer herders' base, but that time I was going very fast, back and forth. On the base, I stayed for just 10 min until I found a reindeer breeder, who said that all had gone for firewood and the cross-country vehicle had also gone somewhere. I went back to the camp and in some 10 or 12 min before the arrival, I heard the helicopter engine noise from the camp and broke into a run, and in 5 or 7 min before my return, the helicopter took off. I was sure that the helicopter came in vain, and my wife and son were left in the camp, because we had just started to pack our things, believing that the flight was to be the next day. So, I was amazed when I saw the empty camp with only my packed rucksack lying alone in the tent. That is, if I had run back a bit earlier, I would only have to take my rucksack and get into the helicopter.

Next day, I left the tundra by cross-country vehicle and started recollecting the amazing components of the situation. The cross-country vehicle driver did as he promised; having arrived at the base he called the airport dispatcher and suddenly learned that the helicopter was to arrive the next day (i.e., a day earlier than planned) approximately around 14.00. This change was caused by the coming several days of military exercises. The driver planned to warn us in the morning, but he was suddenly grabbed by a strong attack of the chronic backache and he could not get out of bed in the morning. Only in the afternoon his fellows found the medicine and made an injection, so he was able to get up and drove to the lake after 15.00. In the morning, the weather was nonflying, windy, and rainy, and began to improve only after 14.00. The driver hoped that due to bad weather the helicopter would arrive not until 20.00, so he would have time to warn us. The cross-country vehicle arrived at the lake camp at 16.50 (20 min after I had left). I had gone far enough and the wind was blowing, so I did not hear its arrival. The driver warned my family about the soon arrival of the helicopter and advised them to pack things quickly, hoping that I would manage to get back in time. Otherwise, he promised them to take me away from the tundra. My wife and son packed the things in a terrible hurry, grabbing the last things when the helicopter had already arrived. They flew off in 5–7 min before my return.

I believe that the random combination of the described factors is impossible. The moments of my departure and return seems to be predetermined. Had I left a little earlier, I would have managed to get to the helicopter, a little later—I would have heard the cross-country vehicle and returned to the camp. The arrivals of the cross-country vehicle and helicopter are also firmly connected. Had the former arrived earlier, I would have heard it and returned back, and later—my wife and son would not have time to pack things. And, on the contrary, if the helicopter had come a bit earlier, my wife and son would not have time to pack things, and if a little later, I would have had time to return. There was also a matched delay in the start time of the actions of all participants. I planned to leave the camp at 17.00 and then changed my intentions. The driver's delay was due to the sudden malady. The helicopter was delayed till evening because of the nonflying weather. It turns out that for the accomplishment of this situation, it was necessary that my actions, the actions of the driver and the pilot, as well as all factors delaying those actions were tightly synchronized. Another distinctive feature of this situation is that it produced a huge emotional impact on all of its participants (my family, the driver, and even the pilot since he was very anxious because of my absence and was ready to follow me to the reindeer herders' base, but no one knew its coordinates).

This situation can be logically explained only if we assume that it was specially made by the tundra nature, while people (participants of the situation) were a part of the tundra, got the signals to begin their actions and obeyed them. Only nature could reconcile the delay of the actions by means of the weather and illness, as well as give the participants signals to begin the action. I clearly felt that signal and obeyed it, as obviously the driver and the pilot did the same, so that the situation was accomplished. Such an explanation of the situation causes the questions: what was the subject of the tundra nature actions and why was this action committed.

The single spiritual principle in nature, approved by the organic worldview, does not mean the universal unity only. The famous Russian philosopher S. Frank wrote: «The world is not a dead machine or the chaos of inert matter, "not a cast, not a soulless mask"; the world is a great living being and at the same time the unity of the multitude of living forces (Frank 1994). That is, the spiritual unity of the world contains spiritual formations of its individual parts. D. Andreev, a Russian prophet-philosopher, beheld these spiritual formations in revelations and called them elementals. Elementals are, in the most general sense, «the elemental forces of the kingdom of Nature» (Andreev 1992). These natural forces are not abstract; they are connected to certain natural forms and have a spatial localization. They appear as living spiritual beings associated with specific natural objects and supporting them. D. Andreev gives many examples of the elementals: those of the bogs, swamps, jungle, desert, deep sea, minerals, the individual mountains, waterfalls, springs, tracts, roots and seeds, grasses and shrubs, trees, storms, hurricanes, hoarfrost, falling snow, fresh snow, eternal ice, blizzards, snowstorms, wind, meadows, fields, rivers, streams; elementals, creating clouds, rain, dew and fog; elementals of the specific places, etc. (Andreev 1992). According to D. Andreev, all natural formations and phenomena are associated with spiritual beings, giving them life.

It is possible to feel the presence of these spiritual beings, according to D. Andreev, in certain circumstances and a certain mood. Various elementals cause different feelings: some—a feeling of fear, anxiety, and danger (elementals of bogs, swamps, jungle, deep sea, etc.), others—a feeling of joy, delight, and admiration (elementals of individual mountains, waterfalls, springs, fresh snow, etc.). Many people are familiar with such feelings in the contemplation of various natural objects and phenomena; the feelings are usually associated with traditional attitudes toward these objects or with their appearance, but according to D. Andreev, the feelings are caused by the contact of a human soul with elementals. D. Andreev gives elementals personality traits and subjectivity by attributing to them the will and ability to interact with humans that is well illustrated by his own interaction with the elemental of the river (Andreev 1992).

Andreev's conception of elementals is well suited to explain the situation described above. I have repeatedly had to experience similar situations of social interaction with nature, where the spiritual formations of various natural objects served as the subjects of this interaction. Several of these situations were described in the previous publication (Tsylev 2016a). Here is a brief description of one of them, which is the most impressive.

I am a true lover of the northern nature and try to go on a fishing trip each summer (at least once a year) to the distant lake, situated in 50–60 km from the nearest settlement, and spend there 7–10 days. There I have one place that is unique in its beauty and convenience of location. I catch lake fish usually by using «torpedka» while floating on a rubber boat. «Torpedka» is a small catamaran (about 30 by 20 cm), which is towed by the boat, not exactly in the stern, but a little aside. Besides the main leash, by which «torpedka» is towed by the boat, there are shorter leashes with baits for fish (Bull trout and Perch). In the described case, I had been sailing for 3 or 4 h, and there had not been a single bite. I was about to return to my camp and sailed along the coast on my way back. Near the coast somewhere ahead, there was supposed to be a sunken plane (since the wartime). Its warped wreckage usually stuck out of the water almost 1 m high. At that moment, for some reason, I thought that the plane was still far and was probably entirely underwater (the water level was very high that year), and I rowed without looking forward (sitting back to front). Suddenly, there was the long-awaited bite! I started to turn, keeping the tension of the main leash to pull the fish, and then saw that I was moving directly to the sticking fragments of the plane, to which only 20–30 m were left. And if not for the fish pecked, I would certainly have stumbled across them and rent the boat.

Flustered, I pulled the fish and was surprised to see that it was hooked by the lower jaw on the outside. That is, the fish did not peck the bait, but was hooked outside as if someone had done it on purpose to prevent my collision with the debris of the aircraft. Such happy coincidences are very difficult to believe. It is more logical to assume it a social action made to me. And the subject of that social action was some part of Nature, most likely, Lake, with whom I was in «a good relationship».

All other cases of my interaction with Nature were also limited to the direct interaction with some natural object. But the tundra case is much more extensive and impressive. Nature was able to synchronize the actions of different people and revealed its capabilities and power. All cases of social interaction with Nature left vivid emotional impressions, but the case in the tundra just brought me in emotional shock because of the incredibility of what happened. Apparently, the subject of social interaction was some major part of the tundra area, so the magnitude of the contact was so impressive. The meaning of that action can be easily interpreted, though it would hardly be possible to find the confirmation.

The most likely are the following three, which are within the context of my tundra studies. The main goal of my research is the study of social interaction between human and nature of the tundra, so the first, most important point is that this very situation as a whole was a vivid demonstration of the enormous possibilities of nature in the commitment of social actions. Nature clearly declared its subjectivity and contact-ability. Second, returning from the tundra by cross-country vehicle alone, without family, I was able to better communicate with the reindeer herders, to observe their life that led me to a conclusion about their involvement in the tundra nature, manifested in their regular orientations to it in their everyday life. And, third, I can add the importance of updating contacts with the herders on my way back. So from the tundra research perspective, my cross-country vehicle trip was very useful.

In this article actions committed by tundra Nature are called social because they contain the basic components of the social action, being directed at a person and having a meaning. In one of our works, we carry out the analysis of the concepts of social and social action, showing that now their content is changing and the scope is expanding greatly (Tsylev 2016b). This is due to the fact that there are more and more evidences of animal behavior, which can be regarded as social. In foreign sociology, there appeared a new term of «extra-human socialite». We all are familiar with the behavior of the domestic animals, when they actually become "members" of the family and build their behavior with a focus on all the other members, but there is growing evidence of such behavior among wild animals. In Russia, this phenomenon is studied by I. Shmerlina, who expanded Weber's concept of social action by including into its potential subjects the representatives of the animal world having a nervous system (Shmerlina 2013).

We made an attempt to expand the concept even more (including nature as a whole and its individual spiritual formations), that is possible with the position of the organic worldview (Tsylev 2016a). Since man is an integral part of nature, the development of human sociality is at the same time the development of nature's sociality, through man nature acquires the features of sociality and becomes their bearer. Therefore, the features of sociality observed in the behavior of animals (and various natural forms and phenomena) are the manifestation of nature's sociality, which appeared because of the development of human sociality.

Thus, the considered cases of human interaction with tundra nature show that, in some situations, a person living in the tundra acts as its integral part. Such situations are well explained with the assistance of the methodology of the organic worldview, which recognizes the existence of the Single spiritual principle and nature's individual spiritual formations able to act as subjects of social interaction and to influence man as part of the One. In the perspective of this approach, nature appears to have all the features inherent in man and can manifest them in suitable conditions. And man becomes the nature's part that can be successfully controlled, although the limits of the control are hard to define clearly. They may depend on the specific situation of interaction. For example, a person in the tundra (or some other part of the «wild» nature, not transformed by human activities and situated in a considerable distance from settlements) largely falls into dependence on nature. Probably, in such cases, man is more susceptible to the influence of Nature and needs the ability to accept its signals, necessary for survival. In conditions of urban life, people are actually separated from nature, appear to be less dependent and are almost not able to feel nature.

Principal Conclusions on the Relation of «Nature» and «Man» Concepts

On basis of the fulfilled analysis, let us turn back to the relation of «nature» and «man» concepts. If nature is associated only with the part of the world's life, which has a material (physiological) basis, while man is associated with all his activities due to social relations, such a Bergersocial» man is already largely rooted in nature. All the social features of human life that are interiorized and have a physiological basis reflected in brain morphostructure, in fact, can be regarded as a natural characteristic, referred to the nature's properties. Purely human (not related to nature) is only that part of man's life, which is not yet assimilated and is performed under conscious control. Moreover, even if some social action is not yet assimilated, but there are some inclinations facilitating the process of assimilation, the action can be considered (at least partially) naturally caused. That means that inclinations are something between the human and natural spheres. The direct heritage of man is only the part of social actions, which is performed under conscious control and which does not yet have inclinations, originating from cultural-historical development of his ancestors. The boundaries between nature and man have a relative character and are held at the level of each individual because each person has a set of inherited inclinations and a complex of social actions and abilities to assimilate.

But from the organic worldview perspective, the separation of the "nature" and "man" concepts can be fulfilled without the orientation on the individuals. Those social actions and abilities that were assimilated, by one person at least, belong to Nature as the whole, as well as all the man's inclinations are the content of Nature. Therefore, in recognition of the nature as alive, it becomes logical to assume the possibility of subjective (or even rational) behavior of Nature as the whole and its individual spiritual formations, as it has been demonstrated in the above examples of interaction with Nature. And purely human, perhaps, can only be called the social actions and abilities that have not yet been assimilated by any person and do not have any inclinations. These are some new actions and abilities that a person is just starting to develop and in the process of their mastering again will give them to Nature. That is, from the standpoint of the organic worldview, man mostly belong to Nature, and only a very small part of him, that is mastering something new for the humanity, remains purely human.

References

Andreev D (1992) The rose of the world. Russian publication: Андреев, Д.Л. Роза мира [Текст]/ Д.Л. Андреев. – М..: Иной Мир. 1992. - 576 с

Berdyaev N (1994) Freedom and the spirit. Russian publication: Бердяев, Н.А. Философия свободного духа [Текст] / Н.А. Бердяев. – М.: Республика, 1994. - 480 с

Berger PL, Luckmann T (1995) The social construction of reality. Russian publication: Бергер, П. Социальное конструирование реальности. Трактат по социологии знания [Текст] / П. Бергер, Т. Лукман; перевод Е. Руткевич. - М.: Изд-во «Медиум» , 1995. - 323 с

Frank S (1994) The meaning of life. Russian publication: Франк, С.Л. Смысл жизни [Текст] / С.Л. Франк // Смысл жизни: Антология./Сост., общ. ред., предисл. и прим. Н.К. Гаврюшина. - М.: Издательская группа «Прогресс-Культура» , 1994. - С. 489–583

Kant I (1995) Groundwork for the metaphysics of moral. Russian publication: Кант, И. Основы метафизики нравственности [Текст] / И. Кант // Кант И. Основы метафизики нравственности. Критика практического разума. Метафизика нравов. - СПб.: Наука, 1995. - 528 с

Marx K (1971) Economic and philosophic manuscripts of 1844. Russian publication: Маркс, К. Экономическо-философские рукописи 1844 года [Текст] / К. Маркс // Маркс К., Энгельс Ф. Соч.: в 50 т.: 2-е изд. Т. 42. – М., 1974. - С. 41–174

Rubinstein SL (1957) Being and consciousness. Russian publication: Рубинштейн, С.Л. Бытие и сознание [Текст] / С.Л. Рубинштейн. - М.: Изд. Академии наук СССР, 1957. - 328 с
Rubinstein SL (1998) Fundamentals of general psychology. Russian publication: Рубинштейн, С.Л. Основы общей психологии [Текст] / С.Л. Рубинштейн. – СПб.: Питер Ком, 1998. - 688 с
Shmerlina IA (2013) Biological aspects of sociality: essays on natural preconditions of human social behavior. Russian publication: Шмерлина, И.А. Биологические грани социальности: Очерки о природных предпосылках социального поведения человека [Текст] / И. А. Шмерлина. – М.: Книжный дом «ЛИБРОКОМ» . 2013. - 200 с
Tsylev VR (2016) Nature as a subject of social interaction: the problem of delimination of sociality and subjectivity. Russian publication: Цылев, В.Р. Природа как субъект социального взаимодействия: проблема определения границ социальности и субъектности [Текст] / В. Р. Цылев // Человек и общество: опыт и перспективы социологических исследований: сборник научных статей / науч. ред. Г.В. Жигунова. - Мурманск: МГГУ, 2016. - С. 172–185
Tsylev VR (2016) The natural and social: the problem of demarcation. Russian publication: Цылев, В.Р. Природное и социальное: проблема разграничения [Текст] / В.Р. Цылев // Воображаемые и реальные миры и их границы: сборник научных статей / науч. ред. А.А. Сауткин. - Мурманск: МАГУ, 2016. - С. 51–61
Weber M (1996) Basic sociological concepts. Russian publication: Вебер, М. Основные социологические понятия [Текст] / М. Вебер // Вебер М. Западноевропейская социология XIX - начала XX веков. - М., 1996. - С. 455–491

Borderology and Practical Knowledge—Humanities Response to Epigenetics?

Jan Selmer Methi

Abstract The Green belt is a good example of what happens in a border zone when the nature is left alone without any disturbance from human activity. What happens is determined by the species' inner potentialities and the living conditions in the belt. The same can be said about development of any organism in general. It is not the potentiality within the organism that determines the result, but the interaction between the potentiality in the organism and the environment. This article will try to show how research on two so different topics as epigenetics and practical knowledge/borderology can benefit from each other's results. Both focus on the condition for interaction in their specific type of border zone.

Keywords Epigenetics · Epigenesis · Practical knowledge · Borderology Border zone · Human activity · Interaction · Living condition · Knowledge

Introduction

At the end of the last century, two new research fields emerged in the academic world; practical knowledge and epigenetics.[1] Researches focusing on practical knowledge saw that there was a contradiction between professional education and professional practice. Many professionally educated people experienced a shock when they first met their practice field. "Practice-shock" later became a term. This problem had been known and looked into for a long time, but the focus on the learning process that the practitioners go through, and the *knowledge* that they gain from the practice field in the process, had not so very much been systematically studied as a separate research field. In 1997, practical knowledge became the

[1]This does not mean that research has not been conducted on these topics earlier. On the contrary, earlier research led to the establishment of the topics as research fields.

J. S. Methi (✉)
Nord University, Bodø, Norway
e-mail: jan.methi@nord.no

J. S. Methi et al. (eds.), *Borderology: Cross-disciplinary Insights from the Border Zone*, Springer Geography,
https://doi.org/10.1007/978-3-319-99392-8_11

research focus for a small group of teachers at Bodø University College. Later, they organized Center for Practical Knowledge and built up master programs and a PhD program within this field.

A completely different topic also emerged on the scene. This field was called epigenetics.[2] According to Wikipedia, the word epigenetics occurs as a term around 1940.[3] Epigenetics shows that not only what is written in the DNA can be inherited. Also, the way grandparents lived (ate, worked with, and were interested in) can have consequences for later generation (Enerstvedt 2011). At the end of the twentieth century, epigenetics was studied from many different angles and perspectives showing that the interaction between genes and its environments determine the ability of the genes to express themselves as phenotypes. What became clear was that a gene's ability to express itself, as do to the interaction conditions in the gene's environment, could also be transferred to next generations. This was new, and if it was true, it would have a great impact on the understanding of inheritance as well as for the concept of learning.

What can research on so different fields, as microbiology and social sciences/philosophy have in common? What are the similarities and what are the differences? Is there any connection at all? These are the problems that this article tries to throw some light into. This topic is in general related to the very old and ongoing philosophical debate which seek to explain the development of individual organic form—epigenesis or preformation—and can be followed from Aristotle in his *Generation of Animals* (1979) through Kant's § 27 in his *Critic of Pure Reason* (1998) to Catherine Malabou's attempt to exceed the dichotomic understanding in *Before Tomorrow: Epigenesis and Rationality* (2016).

In the following text, I will give a presentation of epigenetics in biology and practical knowledge's/borderology's own epigenesis. At the end, I will discuss if there is a connection between them, what the difference is and follow-up Malabou's discussion on what philosophy can learn from science.

Epigenetics

Epigenetics as a research field had to fight its way through the prevailing understanding of genetics to be recognized as an important research field within biology. The results from epigenetics research challenged and maybe even undermined the conceived law of inheritance that only what has been written in the DNA can be inherited. Studies within epigenetics show that changes in the gene expression inherited in the cell line occur without changes in the DNA sequence.[4] Barbara McClintock did a groundbreaking work on «jumping genes» (transportable

---

[2]Epi from Greek: επί- over, outside of, around.

[3]https://en.wikipedia.org/wiki/Epigenetics#cite_ref-10.

[4]This is Lars Olov Bygren's definition of epigenetic (Enerstvedt 2011).

elements TE) in corn.[5] Newer studies show how genes relate to their surroundings, and what these surroundings mean to what extent genes will be expressed or not, whether they should be expressed "weakly" or "strongly", and turned on or off (Enerstvedt 2011).

When Professor Regi Th. Enerstvedt started his basic work on *human activity* as the basic category for all social sciences, he was very much concerned about heredity and environment; generation, genetic inheritance, social inheritance, etc. In one of his books, *Mennesket—kjønn og personlighet* (1979),[6] he says this: "Phenotype in the broadest sense manifests itself in and through human activities on the basis of genotype. The genotype—genes in its entirety—is not potential properties but necessary internal conditions for development of the phenotype" (Enerstvedt 1979, p. 64). As we can see, he is quite close to an understanding of epigenetics, but as he expresses it many years later: "I never asked the question (…) like Lars Olav Bygren did" about phenotypic transfers (Enerstvedt 2011, p. 182).

Lars Olav Bygren, a Swedish professor in social medicine, was one of the founders of epigenetics as a research field. In the 1980s, he started a research project on health in Norway and Sweden. Together with two colleagues, he collected data from generations using historical sources to check the informants' parents' and grandparents' ways of living. These data could relate to marriage, occupation, number of children, properties, reading skills and whatever one could possibly gather from statistics; food, prices, etc. Bygren then asked the following question: Could what parents and grandparents experienced as young persons in anyway be transferred to the next generation? The established science within genetics got a "shock" from the answers Bygren et al. had found in these statistics. Many things seemed to be transferred to the next generation. For instance, abundance and extreme scarcity of food in earlier generations seem to influence the mortality of the coming generation in the family line (Ibid., p. 184). Many studies made afterward tell the same story. Tests on worms fed with a special germ indicate that changes in phenotype can last for 40 generations without changes in genotype (DNA) (Ibid.). Scientists are still skeptical about drawing a solid conclusion, but what can be said is that the studies on epigenetics seem to show that "(….) gene expression, i.e., phenotype, *can be transferred between generations*" (ibid.).[7]

An interesting article by Roar Fosse on mental illness, heredity, and environment in a Norwegian journal on psychology published in 2009, presents good arguments against linking genes and mental illness. He refers to recent years' research on DNA that shows that it is not the gene itself, but the processes in the nucleus around the DNA that determine the expression of the gene. According to Fosse, these epigenetic processes are strongly influenced by environmental experience of the

[5]https://www.nature.com/scitable/topicpage/barbara-mcclintock-and-the-discovery-of-jumping-34083.

[6]In English *Human—sex and personality.* My translation.

[7]In original language «(…) genuttrykk, dvs. trekk ved fenotypen, *kan overføres mellom generasjoner".*

individual organism through childhood and adolescence. Mental disorders seem to evolve when experiences from a stressful environment modifies epigenetic processes, and consequently, both brain development and psychological functions into adulthood. This must, in his opinion, lead to a different view on the causes of psychological difficulties (Fosse 2009, p. 596). In his article *Genetikk—Er betydelige effekter dokumentert ved schizofreni?*,[8] he concludes that a number of DNA variants are associated with schizophrenia, but the explained variance is modest (Fosse 2017a, p. 455). In his article *Epigenetiske endringer ved alvorlige psykiske lidelser—et resultat av psykososialt stress?*,[9] he promotes the hypothesis that stress-induced epigenetic changes play an important role in serious mental suffering (Fosse 2017b, p. 941). His conclusion is: "Epigenetic changes in severe psychiatric disorders are a new area of reflection that has not yet been oriented towards a psychosocial stress perspective. It remains to be investigated whether epigenetic changes in severe mental disorders coincide with those seen after psycho-social stress in early life" (Ibid., p. 947). Reading his article, one can say that there are strong indications, based on animal research, that psycho-social stress early in life can change CG methylation of genes in hypothalamus, hippocampus, and prefrontal cortex. CG methylation usually alters how genes are expressed. According to Fosse, postmortem studies on humans show very much the same. The article also shows, discussing the challenges with the previous research, that even if CG methylation "considered to be the most stable epigenetic marker, it is still relatively dynamic and changes significantly from adolescence to adult life and old age" (Ibid.).

In general, one can say that research on epigenetics shows that it is the genes' interaction with the conditions surrounding the genes that determine if a gene is turned on or off or is expressed weakly or strongly. It can also be said that this has consequences for the development of the whole organism and that there are some indications that some parts can be changed throughout lifetime of the organism.

Enerstvedt takes an even stronger position in contradistinction to the standard view, as it was then, and even now, when epigenetics is officially included in biology as teached. I tend to agree with Enerstvedt, there is certainly an inflation in making the genes, understood as sub-molecular structures of the DNA molecule, themselves actors, even if only potentially. The potentialities that are linked to these molecular structures are essentially connected to manners in which they are orchestrated to function, prima facie in the cells, and then basically in the bodies themselves. And as Enerstvedt many times underlines: "The essence of life is characterized as the interaction between organism – surroundings..", and "the genes are therefore not the unit of the selection" (Enerstvedt 2011, p. 179).[10]

[8]In English *Genetics—Are substantial effect documented in schizophrenia?*.

[9]In English *Epigenetic changes in severe mental illness: a result of psychological stress?*.

[10]In Norwegian: «Livets vesen er karakterisert som vekselvirkningen mellom organisme – omgivelser», og «Genene er derfor ikke seleksjonens (utvalgets) enhet.» This is a part of Enertvedts critick of Richar Dawkin's *The Selfish Gene* (1976).

Practical Knowledge and Borderology

In this part, I feel it is necessary to take a historical detour for the presentation of the research field of practical knowledge to show that practical knowledge can be seen not only as an analogy to epigenetics, but also as a part of the epigenesis of knowledges in general.

In the European and western tradition, theoretical knowledge has been in focus for more than 2000 years. With references to Plato and Aristotle, this tradition has focused on inference as the safest deriving form of knowledge. Inferences are steps in reasoning, moving from premises to conclusions. Mainly, two types of inference are being used. Deduction is an inference deriving logical conclusions from premises known or assumed to be true. Induction is inference from particular premises to a universal conclusion, today usually backed up by statistical methods of verification. In this tradition, theoretical knowledge is the safest form of knowledge, and philosophical studies of knowledge, in general, is hence still called epistemology, in accordance with Aristotle's term coined to be characterized in the intellectual duty connected to the knowledge for *theoria, episteme.*[11]

The search for true knowledge within the form of practical knowledge has a shorter history, but aside from earlier attempts, even going back to Plato and Aristotle, most of it is rather a recent date. In the publication *Praktisk kunnskap—som erfaring og som forskningfelt* (Nergård et al. 2005),[12] the changes started in 1800/1900, especially with the American philosopher William James and his presentation of Charles Sander Peirce's thinking. One might say that the American Pragmatism was born. In the article *How to Make Our Ideas Clear* from the 1870s, Peirce presents his thesis that theoretical concepts that have no practical consequences are empty. This was a big step in a new direction. The founder of the modern working life research, Frederic Winslow Taylor published his essay *The principles of Scientific Management* in 1911. Here, he sums up the systematic research on physical hard labor that had been conducted for more than 20 years. Taylor shows that at least some parts of working life can be studied scientifically. His vision was that all aspects of working life can be explored scientifically. The American pragmatism was influenced by Kant's attempt to analyze theoretical cognition as based in the knower's practical experience, constituted through acts on different levels. Kant's transcendental philosophy can be said to reactivate a motive in Greek philosophy in general that knowledge (*episteme*, *techne* and *phronesis*, according to Aristotle) is (intellectual) virtue. This implicates that we cannot reach knowledge without experience or conceptual work or some kind of linguistic practice. The HBO-report, an essential document in the establishing of practical

---

[11]Episteme; from Greek ἐπιστήμη, *epistēmē*, meaning "knowledge", and λόγος, *logos*, meaning "logical discourse". Wikipedia: https://en.wikipedia.org/wiki/Epistemology.

[12]In English *Practical knowledge—as experience and as research field.* Hereafter referred to as the HBO-report. The paragraphs below are based on information from the HBO-report. Otherwise reference is given.

knowledge as the research field upholding the PhD degree "Studies of professional practica" at Nord university, also refers to Gilbert Ryle's distinction between "knowing that" and "knowing how" (1949), Ludwig Wittgenstein's ideas about thought and language (1930/53), Michael Polanyi's work on "personal knowledge" and "tacit knowledge" and the Norwegian philosopher Jakob Meloe's very personal further development of impulses from Wittgenstein and Heidegger, often referred to as "praxeology".

One other decisive factor that had a great impact on the study of practical knowledge, was the emergence of applied ethics in the beginning of 1970s. One of the pioneers within the field, Stephen Toulmin, summarized the situation in his article, published in 1973, *How Medicine Saved the Life of Ethics*. An overview of the field practical knowledge from the last decades of the twentieth century, came with Donald A. Schön's book *The Reflective Practitioner* (1983), and Bengt Molander with his book *Kunnskap i handling* (1993).[13]

In Scandinavia, some small environments were working in the same direction. In Denmark, Uffe Juul Jensen's book «*Sygdomsbegreper i praksis. Det kliniske arbejdes filosofi og videnskabpsteori*»[14] (1983) made a great impact on the understanding of what nurses and doctors do and should do. The same happened with Patricia Benner's book *From Novice to Expert. Excellence and Power in Clinical Nursing Practice* (1984). In the environment evolving around Bo Göranzon and Ingela Josefson in Stockholm, Sweden, a systematic work on experience and practice was initiated. Göranzon formed a research field called *Yrkeskunnande og teknologi*.[15] Impulses inspired by Wittgenstein and mediated by philosophers from Bergen, Norway, soon became a part of public debate. Among concepts used were "propositional knowledge", "skill knowledge", and "familiarity knowledge". In Norway, Kjell S. Johannessen, Tore Nordenstam and Gunnar Danbolt published the book *Den estetiske praksis* (1979).[16] This was the product of a cooperation philosophers and art and theater historians. Later, Nordenstam's book *Fra kunst til vitenskap* (1987)[17] also became an important contribution to the field of practical knowledge. The Norwegian philosophers Jakob Meløe and Viggo Rossvær played a key role in the field for more than three decades, focusing on human activities, especially within the Norwegian coast culture and fisheries. Meløe's article *Words and object* shows that Meløe is developing a deep understanding of the concept of human activity.[18] Furthermore, Rossvær's book *Ruinlandskap og modernitet. Hverdagsbilder og randsoneerfaringer fra et nord-*

[13]In English *Knowledge in action.*

[14]In English *Illness Concepts in practice. The clinical work's philosophy and philosophy of science.*

[15]In English *Working Skills and Technology.*

[16]In English *The esthetical practice.*

[17]In English *From art to science.*

[18]This category is central in the Activity theory. In Norwegian, the concept is *virksomhet,* in Russian *Dejatelnost* and in German *Tätigkeit.*

norsk fiskevær (1998),[19] is an important contribution to research on practical knowledge. This book shows the close, almost symbiotic, and relationship between man and nature.

The systematic and institutionalized research on professional practice did not start in Scandinavia before 1990. A research group called *Sundhed, menneske og kultur*[20] at Århus university, led by Uffe Juul Jensen, started to look at professional practice from three different angles; the professional practitioner's, the patient's, and the Danish welfare state's. The research group later became a research center with the same name.[21]

Based on the knowledge referred to above, and the fact that professional education almost had lost its connection to the practice field, Bodø University College established a Center for Professional Studies in 1997, later called the Center for Practical Knowledge (2001). In 2000, the center could offer a Master in professional studies, later called the Master in Practical knowledge (2002). The basic contention built on was that practitioners acquire knowledge through their working activity that could not be acquired otherwise, and that is difficult or sometimes impossible to articulate and communicate adequately in words. This knowledge emerges through the action (knowledge in action). The master program offered practitioners the opportunity to write forward their own working experience, reflect on it and develop new knowledge from the experience.

In 2009, Bodø University College was licensed to offer the first PhD program in practical knowledge. Almost at the same time, and with the Center for practical knowledge in Bodø as model, a similar center was established in Stockholm at Södertörn University College. With these institutions and their educational and research programs, one can say that practical knowledge as science and research field had finally been approached systematically, even though the pioneers had opened the field more than a century ago.

In 2012, Nord University and Murmansk Arctic State University established a Joint Master Program related to practical knowledge. This program was called the Joint Master Program in Borderology. The education program was institutionally located in the border zone between Norway and Russia.

If one looks at the presentation in the HBO-report more closely, one will notice that there are many blind spots both in the understanding of the concept and int the identification of research areas on practical knowledge. However, in a short report , one cannot expect to find a comprehensive study of the field. But the report shows in what direction the authors are moving. The process of building up a new subject for research was a very complicated and tough task to tackle for the small environment in Bodø. To establish a PhD in and well known and internationally

[19]In English *Ruin landscape and modernity. Everyday pictures and rand zone experiences from a Northern Norwegian fishing village.*

[20]In English *Health, Human and Culture.*

[21]Information from the HBO-rapport 10/2005 pp. 8–10 stops here.

recognized field is difficult in itself. To do so within a field that is not yet recognized is something completely different and much more difficult.

In the following, I will present an approach to the concept of practical knowledge from a different perspective connected to the founders of Activity theory and Cultural Historical School. Their philosophical background was dialectical materialism, focusing especially on how Karl Marx had analyzed capitalism, and his criticism of Ludwig Feuerbach. Marx' *Thesis On Feuerbach* (1845)[22] became fundamental for this school since he places human beings in dialectical interaction with the environment (nature) and thus shows that man changes himself at the same time as he is processing the nature outside himself. This thought was evolved and put to use in the theory of learning by a scientific group in Leningrad in the early 1920s. Here we find L. Vygotskij, A. Luria and A. Leontjev as leading figures. The legacy of this group is today managed by the organization ISCAR.[23] The founders of Activity theory questioned the dichotomous understanding of environment and heredity and stated a new way of looking at the relationship. With support from their developed insight and scientific work, they proclaimed that the special human psychic abilities and activities must be developed through and only through human activity (Enerstvedt 1982).

In Western Europe, Activity theory was introduced as late as in the mid 1970s. The first international conference on the theory was held in West Berlin in 1986. Representatives from 20 countries participated (Broch 1991, p. 11). Since then, the ISCAR conferences have been held regularly.

In Scandinavia, the two strongest environments for Activity theory have been in Denmark and Finland. An interesting point here is that Uffe Juul Jensen, mentioned above, has been a central figure both for the introducing of Activity Theory and practical knowledge. There is also a special connection between Jakob Meløe and Uffe Juul Jensen. Jensen was a student in philosophy in Copenhagen when Meløe introduced the work of Wittgenstein to the students. In Finland, Yrjö Engeström (2009) with his Activity System and the concept of expansive learning has been a leading figure. In my opinion, the most important representative of Activity theory, and the one with the deepest understanding of the basic philosophical principles of both practical knowledge and Activity theory has been the Norwegian social scientist and philosopher Regi Th Enerstvedt (1938–2017). With his always reflecting attitude towards his own work and the work of others, he has been able to deepen the understanding of the concept of human activity. We can see that through his comprehensive production, represented in particular by two books; *Mennesket som virksomhet* (1982) and *Spiller Gud terning likevel?* (2011).[24] In his latest article *Kontroversen mellom Vygotskij-skolen og Rubinstein-tradisjonen*,[25] he develops an

[22]Written in 1845 but first published as an appendix to *Ludwig Feuerbach and the End of Classical German Philosophy*, 1888.

[23]The International Society of Cultural-historical Activity Research (ISCAR).

[24]In English *Human as activity* and *Does God play dice anyway?*

[25]Tidsskrift for Norsk psykologforening, Vol 53, nr. 9, 2016, pp. 740–743.

even deeper understanding of the relationship between internalization and externalization when it comes to learning and human activity. His understanding, based on the basic categories in Activity theory, establishes the link between Activity theory and epigenetics. Enerstvedt is explicit on this link when he says: "Epigenetics is a further argument for my claim for over 30 years that all human's distinctive mental qualities (…) are formed in, and only in, the activity" (Enetrstvedt 2011, p. 185). But we can also argue that within the understanding of Vygotskij's famous concept *the zone of proximal development* lays a deeper insight that it is the conditions within the zone that determine what is being developed as solid or weak. His work, based on practical experience with disable children, shows that (Vygotskij 2004, p. 279).

The professional community at Nord university, which can be considered as the leading environment for studies on practical knowledge, has published four volumes on practical knowledge and borderology since 2015. The first book that came from the group, attempting to answer the question about the core of practical knowledge as science and research field, was published in Norway in 2015. The book was called *Praktisk kunnskap som profesjonsforskning* (McGuirk and Methi 2015).[26] This anthology presents practical knowledge from different perspectives but does not go deeply into the question of what makes the subject of practical knowledge such an exiting new field of research, and what is so special about the concept practical knowledge. An honorable attempt is Anders Lindseth's definition proposing that practical knowledge is the *ability to answer* (Lindseth 2015).[27] This does not only refer to a verbal answer, it is the whole body's ability to answer. A gesture is also an answer, a response that is interpreted as an answer. To this, I will say that it is a challenge to understand practical knowledge as an ability to give a response (give an answer) to something. One problem here can be a hidden maintenance of a dualistic view that we, on the one hand, have practical knowledge and on the other theoretical knowledge. It will break the frame of this article to discuss Lindseth's definition on practical knowledge, so I will leave it here. I will only state that practical knowledge is something that is established as resulting from all (kinds of) human activity. This indicates that human beings gain insight into circumstances (subject matters) through participating in human activity. This again indicates that there is knowledge we can articulate in words, and there is knowledge we can only demonstrate through action. Both types of knowledge are of interest for the research field on practical knowledge.

With the Joint Master Program in Borderology, we discovered that the border zone between two countries can also be used as a concept for the arena where new knowledge is developed. This must be understood both as conscious and unconscious development of knowledge. Unconscious means in borderology context, unarticulated knowledge developed in an environment of more than one culture. This knowledge will be specific for the area, and may not be seen elsewhere.

[26]In English *Practical knowledge as research on professional work experience.*

[27]In original language *praktisk kunnskap er svarevne.*

Conscious means here what happens when one deliberately meet with one another to gain knowledge from each other's experience. In the borderology, program we saw similarities with the zone of proximal development (Vygotskij) used in a pedagogical context. But in contrast to the zone of proximal development where another more skilled pupil or teacher is a central part of the conditions for the development, all human contributors are on the same level in Borderology. None of the participants knows the answer to the problem or the task. The development is construed as a Bakhtinian heterogenously based dialog (interaction) between personalities on an equal level. Viggo Rossvær's article *Den praktiske kunnskapens teori*,[28] shows that the concrete border zone between Norway and Russia, like any other border zone on the globe, offers "a workshop for cosmopolitan communities where both sides contribute to the social relationship to the other, and that of seeing the world from the standpoint of the other" (Rossvær 2015, p. 104).[29] In my article in *The border zone as an arena for exceeding oneself* (Methi 2015), I see the border zone as the place where the activity leads to a result that only can be understood as determined by the activity in the zone. I try to show how students, by using essay as a critical investigation method on their life experience, are able to question the establish truth in the experience, exceed the truth and build up a new and stronger understanding of the circumstances within the same experience. Based on the understanding of Marx' *Thesis on Feurebach* and Hegel's dialectical method (being—nothing—becoming), the activity in the "border zone", used as a general concept, appears to be the core of practical knowledge, and maybe the fundamental ground for understanding the epigenesis of knowledge.

Two aspects can be drawn from this presentation. The first is that the conditions within the border zone surrounding an activity, and the activity in itself determine *what* changes take place, when integrated with each other. Changes are the core for understanding what learning is. Second, the border zone can be deliberately created to exceed an established understanding of a phenomenon. This is what the activity within the research area practical knowledge and borderology is about; to exceed oneself.

Discussion

The analogy between epigenetics and practical knowledge/borderology must of course not be drawn too far. But what we can gather from the presentation above about epigenetic is that it is the relationship between the genes *in the organism* and the environment that determines if and how a gene expresses itself. This has

[28]In English *The theory of practical knowledge*.

[29]In original language «Grensesonen tilbyr her – i likhet med andre grensesoner over hele jordkloden – en workshop for kosmopolitiske felleskap der både det sosiale forholdet til den andre og det å kunne se verden fra den andres standpunkt».

consequences for the development of the organism which again can influence what is heritable in the cell line without any changes in the DNA.

One important distinction is necessary; it is the organism, not the gene that is active. A gene in itself cannot be active. They can only have a basic function when internalized in a living organism. "The gene's information is a component of a system, the organism where the distinctive analog system-based information activity is based on the distinctive analog human phylogenetic information. This distinctive information system characterizes the human as species" (Enerstvedt 2011, p. 178). This is also the basic background for understanding the ability human beings have to form all distinctive human attributes, including freedom as an aspect of human nature.

Another contribution to understanding human beings as practically knowledgeable, indicated by the presentation of epigenetics, is that methylation can change throughout life. This indicates that a methylation, established early in life through certain activities, can be changed as a result of other activities later in life. In other words, what is determined by human activity can be changed by human activities. And again, the determination is not written into the genes. It is the opposite. The determination of the gene's phenotype, how it is expressed, is determined by the activity of the organism, by the way in which the genes are related to its environment of the organism, e.g., the human being.

If we use the border zone as a metaphor and analytically describes it as the arena where the activity and changes take place, then we can say that research on epigenetics shows that nature, in this case the organism, creates its own nano-border zone where the potentiality of possibilities may or may not be real. The development is determined by the different parts that participate in the activity in the border zone. The metaphor must only be understood and used as a picture of what goes on unconsciously.

Then, we can return to practical knowledge and borderology. When a professionally educated person starts his/her professional career, it is not the potentiality of possibilities from the education that shows itself in the professional worker's practice (the appearance of the professional identity), but how the potentiality of possibilities from the education is made possible by the relationship to the working conditions. This means that it is the characteristics of the activity that allows the possibilities of becoming realized or changed. In my research on professional identity, I saw that for many newly educated teachers, the first meeting with the professional field was a shock. Many of their ideas about the work they carried with them from the education were smashed. However, step by step, the interaction between the potentiality of the education and the working condition developed what later became the person's professional identity (Methi 2007). The border zone for them was just there, in the meeting between the professional workers and their practice.

At Center for Practical Knowledge at Nord University in Norway, we let the students study their own experience from their professional work and (life experience) thus creating insight that can be subsequently implemented in the same field. The way we accomplish this, is by creating an arena or a border zone, where the experience can be critically reflected. The critical reflection often leads the students to new, deeper and broader understanding of themselves as professional workers. What we often see is that their first understanding of the truths laying deep within their experience presented in their narratives dissolves and is replaced by a new and stronger understanding. The border zone then becomes an arena for exceeding their established view, not as a denial of former truth, but as a new and exalted truth that preserves the best of the old. In other words, it is possible to consciously establish an activity to overcome a vision that earlier in life has been seen as true.

If we bring in the concept of learning the way Enerstvedt recommend; learning as a change in the system as a part of all activity, and learning as a wanted change in an object-directed activity in order to develop new skills or insight (Enerstvedt 1986), we can see that knowledge has its own epigenesis. The first part of the learning concept follows the track of research within epigenetics. The latter follows the track of epigenesis within philosophy. According to Christopher Watkin (2016), the idea of Catherine Malabou in her book *Before tomorrow: Epigenesis and rationality* (2016) is that "epigenesis offers ut a new way if understanding the Kantian transcendental and a new paradigm of philosophy in general".[30] For us this means that practical knowledge and borderology must incorporate both the epigenesis of knowledge from philosophy and the recent neurobiological discoveries that suggest a profoundly transformed view on brain development which again point to the concepts of learning and knowledge.

Conclusion

How can we say that practical knowledge and borderology is response of the humanities to epigenetics? In my opinion, there is one important parallel that justifies the statement. Research on epigenetics so far, have strongly indicated that the way we live and act can be written into the gene's phenotype and transferred to the next generation. This suggests to us that what is transferred through the phenotype can also be changed by the activity we participate in. Practical knowledge and borderology study the knowledge that is developed through lifeworld activities and working experiences, and by participating in a critical reflecting activity on own experience, the knowledge acquired through the experience can be complemented, corrected, and strengthened or changed.

For both, the result of the studies can implicate that it is the human activity/the activity of the organism that determines what we become and what is inherited without a change in the DNA.

We can at the end suggest that the relationship between epigenetics, epigenesis, and practical knowledge are rooted partly through analogy, partly through identity

[30]The article online has no to pages.

in function, and partly because human development of knowledge "only" is a continuation of the kind of learning that the organisms are generally shown to be capable of through epigenetics.

References

Broch H (red) (1991) Virksomhetsteorien. En innføring. A/S Falken forlag. Oslo

Engeström Y (2009) The future of activity theory. In: Sannino A, Daniels H, Gutiérrez KD (eds) Learning and expanding with activity theory. Cambridge University Press, NewYork

Fosse R (2009) Ingen gener for psykiske lidelser. Tidskrift for Norsk Psykologiforening [online] Volume 46, nr 6, pp 596–600. http://www.psykologtidsskriftet.no/index.php?seks_id=82599&a=4. Accessed 2009

Fosse R (2017a) Genetikk – er betydelige effekter dokumentert ved schizofreni? *Tidskrift for Norsk Psykologiforening* [online] Volume 54, nr 5, pp 454–465. http://www.psykologtidsskriftet.no/index.php?seks_id=485206&a=2. Accessed 3 May 2017

Fosse R (2017b) Epigenetiske endringer ved alvorlige psykiske lidelser. Tidskrift for Norsk Psykologiforening [online] Vol 54, nr 10, 2017, side 940–950. http://www.psykologtidsskriftet.no/index.php?seks_id=497383&a=2. Accesed 5 Oct 2017

Lindseth A (2015) Svarevne og kritisk refleksjon – Hvordan utvikle praktisk kunnskap? In: McGuirk J, Methi JS (ed) Praktisk kunnskap som profesjonsforskning. Fagbokforlaget. Bergen, pp 43–61

Marx K (1845) Thesis on Feurebach. In: Engelsh F (ed) Ludvig Feuerbach and the end of classical German philosophy. International Publisher, New York

McGuirk J, Methi JS (2015) Praktisk kunnskap som fag- og foskningsfelt. In: McGuirk J, Methi JS (eds) Praktisk kunnskap som profesjonsforskning. Fagbokforlaget, Bergen

Methi JS (2007) Psychagogi. Fremmedgjøring eller frigjøring. En studie av lærernes yrkesidentitet og selvforståelse i perspektiv av virksomhetsteorien. PHD i filosofi. Universitetet i Århus

Methi JS (2015) The border zone as an arena for exceeding oneself. In: Rossvær V, Sergeev A (eds) Philosophy in the border zone. Orkana Akademisk, Oslo, pp 213–225

Negård J et al (2005) PhD-utvalget ved Profesjonshøgskoeln. HBO-rapport 10/2005. Bodø

Rossvær V (1998) Ruinlandskap og modernitet. Hverdagsbilder og randsoneerfaringer fra et nord-norsk fiskevær. Spartacus Forlag A/S, Oslo

Rossvær V (2015) Den praktiske kunnskapens teori. In: McGuirk J, Methi JS (ed) Praktisk kunnskap som profesjonsforskning. Fagbokforlaget. Bergen, pp 103–125

Enerstvedt R Th (1979) Mennesket. Kjønn og Personlighet. Forlaget Ny Dag. Oslo

Enerstvedt R Th (1982) Mennesket som virksomhet. Tiden Norsk Forlag. Oslo

Enerstvedt R Th (1986) Hva er læring? Falken Forlag AS. Oslo

Enerstvedt R Th (2011) Spiller Gud terning likevel? Marxist Forlag A/S. Oslo

Vygotskij L (2004) Problemstillinger I undervisningen og den intellektuelle udvikling I skolealderen. In: Gunilla L (ed) Vygotskij om læring som udviklingsvilkår. Århus, pp 267–285

Watkin C (2016) Before tomorrow: epigenesis and rationality by Catherine Malabou. In: Nortre Dame Philosophical reviews, an electronic journal, 223 pp. https://ndpr.nd.edu/news/before-tomorrow-epigenesis-and-rationality-2/

Human Existence Between the Forest (Nature) and Home (Technology)

Vasilii M. Voronov

Abstract In the article views (considered, investigated) the situation in the sphere of human-nature-technology relations in various philosophical-anthropological aspects. The environmental issues are understood as a result of the human existence that, on the one hand, cannot explore the world to the full, and, on the other hand, always strives to have its fully explored place in it. The mental images of 'home' and 'woods' are used by the author as phenomenological and symbolic clues for understanding the essence of the way technology and nature are related to people. The possibility of solving the present-day environmental issues by a discretionary action, for example by adoption of new environmental strategies or by changing of the sociopolitical regime, is put under scrutiny. Methodologically, the author uses the perspective of the phenomenological understanding of the 'living matter' (woods, ὕλη) developed by the famous Russian philosopher and translator V. V. Bibikhin.

Keywords Human · Nature · Environmental problems · Woods (forest) Technology · Home

1. A general idea constantly bringing relevance to the global environmentalist discourse is the troubling character of present-day relations between humans and the nature. It has already been accepted to speak about various challenges and risks of the modern technocratic civilisation. In our view, we can specify four core aspects most represented in the modern global environmentalist 'talking': 1. *already existing environmental issues related to the negative impact of human activities on the biosphere; 2. resource issues related to the environmental issues; 3. alienation of the humans from the nature; 4. challenges and threats related to the risk horizon of technocratic activities of humans.* All the mentioned aspects have a bilateral character since they affect nature with one side and humans with the other.

The aspects should be briefly explained. Relevant environmental issues are those we see right now or can (will be able to) calculate in the nearest future.

V. M. Voronov (✉)
Murmansk Arctic State University, Murmansk, Russia
e-mail: philosophvoronov@gmail.com

J. S. Methi et al. (eds.), *Borderology: Cross-disciplinary Insights from the Border Zone*, Springer Geography,
https://doi.org/10.1007/978-3-319-99392-8_12

Undoubtedly, some of those issues are debatable. It is sufficient to be reminded of the debates about whether the manmade greenhouse effect as a global warming actor is real or false. Undoubtedly, the modern scientific discourse is dominated by the idea that the very fact of the global warming is real and that the manmade greenhouse effect is a contributing factor. For instance, this is shown in conclusions of the IPCC Fourth Assessment Report and IPCC Fifth Assessment Report. Moreover, there are widely known political decisions based on those assumptions, like the Kyoto Protocol ratified by many countries and suggesting control and quotas of greenhouse gas emissions (carbon dioxide, methane, fluorocarbons, nitrous oxide and sulphur hexafluoride) into the atmosphere. At the same time, the very idea of global warming or, at least, its manmade nature is criticised in the public opinion and political circles and scientific community. The USA has notably signed but declined to ratify the Protocol.

The doubts in the global environmental discourse are connected with using (by funding activities) radical environmentalist organisations in global economic competition and geopolitics. In particular, both Russian authorities and the public have a sceptical mood about funding sources of those organisations explained by the mundane logic of ‘who pays calls the tune’. For instance, the known action of Greenpeace activists on board of *Arctic Sunrise* in the Pechora Sea (on the Prirazlomnaya oil platform) was received by the Russian society with suspicion. The version of economic rivalry against the Russian Federation by non-economic means was popular, which can be illustrated by polling data of ‘Russian public of opinion research center’ (VTsIOM): with 60% of the respondents considering the actions by Russia adequate to the situation, 17% said the actions were too harsh and 8% said the actions were too soft[1] (VTsIOM Press release No. 2418). It is important to note that 66% of the respondents thought it would be necessary to thwart future attempts by foreign environmentalist organisations to prevent Russian Arctic exploration, with just 20% of the respondents holding the opposite position. We should note that these events took place before the increase in geopolitical tensions between the RF and the West due to the events in Crimea and Ukraine in 2014. The reaction in the public opinion is therefore unlikely to be tied to a certain situation.

At the same time, doubts and differences in certain environmental issues should not lead to discrediting environmental issues as a whole. Denying the existence of any of these problems as such requires ‘hiding your head in the sand.’

The second aspect of human-nature relations means the actual or possible depletion of the natural components perceived as resources in the modern era. Such depletion and risk of loss may, for example apply to fresh water, oil, minerals, etc. The situation can be very accurately shown by the metaphor of a fire the humankind is sitting around (Bibikhin 2011). While in the primeval era firewood was used,

[1]The VTsIOM poll throughout Russia was done on 28–29 September 2013. 1600 people were polled in 130 inhabited localities in 42 Regions (Oblasts, Krais and Republics) of Russia. The margin of error is 3.4%.

nowadays, the fire of global consumption and therefore energy consumption is kept alive, for example by hydrocarbons or nuclear fuel.

The alienation of the human from the nature is also a complex issue and can manifest itself in worldview priorities, in economic and political decisions and everyday reality of people, with the modern technocracy preparing ways to increase this alienation that earlier could only appear in science fiction. For instance, cybernetisation of the human body or genetic engineering eugenics may well be done in the future. Finally, the modern level of technical development poses threats that are not direct consequences of planned and calculated activity but happen or may happen due to various accidents and circumstances beyond control. The range of these capabilities is very wide, from relatively local man-made disasters to a global nuclear slaughter.

The summarised challenges and risks are often exclusively attributed to the modern civilisation, rooted in a certain worldview shift and a loss of the archaic harmonious relation with the nature. Such reflections view the source of all problems in the situation of alienation from the nature. This worldview is often explained to have begun in the industrial civilisation appearing in the early modern period, gradually spreading around the globe. Bearing the worldview aspect in mind within this understanding of the issue, we need to mention the New European assumption *scientia potentia est* that set a worldview basis for the scientific and technical progress. However, despite this widespread opinion, we can provide well-known examples of impact on the nature by pre-industrial societies. In particular, we can remember: the total deforestation done by the people of the Easter Island, the well-known cliche about 'goats eating the Greek might', the emergence of the Sahara Desert likely caused by economic activities or the contribution of the Clovis prehistoric culture to the destruction of the American megafauna, etc. The problem of relation between the human and the nature are therefore not something that is specific of a technocratic civilisation only. On the contrary, the alienation of the human and the nature may be understood as an essential philosophical anthropological phenomenon requiring an explanation.

2. The complexity of human-nature relation already existing in the primeval era means that the archaic human was already alienated from the nature. Symbolically, it can be expressed by images of *'going out of the forest'* and *'finding a home'*. The *forest*, in this case, is understood by us in the expanded meaning assigned to it in the phenomenological concept of V. V. Bibikhin in relation to the term of the Aristotle's philosophy—*ὕλη* (Bibikhin 2011).

For our purposes, we can specify the following aspects of understanding the phenomenon of the forest by V. V. Bibikhin: *1. Living (organic) matter, 2. Matter per se, 3. Forest element, 4. Female (sexual, maternal) principle, 5. Perfect, automotive automat, 6. Sophia of the world.* The said meaningful horizons are not analytically itemised by Bibikhin, but rather synthetically complement each other. How is it possible? Probably by connection and mutual enrichment of different approaches in theoretical ideas: phenomenology, logical-linguistic interpretation of language, ideas of J. Derrida and Christian philosophy (hesychasm, the Russian tradition of religious philosophy). Bibikhin pays attention to the fact that the very word *materia* (Lat.) is a Latin philosophical term used to translate the Greek philosophical term ὕλη, the first meaning of which is 'woods.' In its turn, in the Latin language, this meaning was likely a secondary one, with the primary understanding being a material, primarily a combustible material, and, first of all, timber, i.e. woods (Bibikhin 2011, pp. 11–16). Consequently, the forest is already etymologically present in any contemporary conversation about matter and the material. On the other hand, this link entered into the philosophical and everyday language due to human experience in interaction with the forest, for instance, as a supplier of energy consumed by people (firewood in a fireplace). At the same time, Bibikhin also notes other, more intimate, sacred, or archaic forms of connection between the human and the forest. An experience of the forest is also hair growing on the body, the symbol of the cross that may be connected to the symbol of the world tree, the sacred practice of relation to trees in the archaic culture. We can pay attention to the intimacy of the archaic relation to the forest as opposed to the utilitarianism of the modern civilisation. The civilisation, according to the Russian philosopher, really pays a price for its attempt to distance itself from the forest. The price is the attraction of a modern urban dweller to its substitutes: 'wine, tobacco, and drugs' (Bibikhin 2011, p. 5). The forest is therefore shown by Bibikhin as a certain principle of being, as an expression of the natural material principle. Moreover, it is specifically shown in the phenomenological sense as analytics of different forms of experiencing the forest or, more exactly, analytics of proximity of and attraction to the forest.

The forest as nature and living matter gives a human a unique experience of a *perfect automat*, i.e. an automotive, self-replicating principle. For example, we can notice this experience in vegetation of plants, reproduction of animals, instinctive behaviour, etc. The *perfect automat* is interpreted by the Russian philosopher as the *'Sophia of the world'* (Bibikhin 2011, p. 93). To explain it here, we can refer to metaphysical constructs of Aristotle, namely, the famous study of four causes. As the cause for existence of things, Aristotle famously specified: form (what), ὕλη-matter (from what), purpose (for what) and agent (source of movement, change or rest) (Aristotle Physics II, 3, 194b16–195b30, Aristotle Metaphysics I, 3, 983a25–983b5). The said causal principles, according to the ancient Greek philosopher, 'work' both in the nature and in human activity but in the nature this link is, in a way, more organic. The nature is aptly compared by Aristotle with 'a man who heals himself' (Aristotle Physics II, 8, 199b30). In the Bibikhin's concept, this self-motion of the nature is interpreted as the wisdom of the world (divine) organisation: '*Sophia of the world*'.

At the same time, Aristotle sees in human activity either an imitation of the nature or creation of something the nature cannot create. In our view, however, the important thing is that the human, by doing so, breaks the automotive causal chains. The human sort of imposes their purposes on other things, adapting and adjusting them for and to the human. In this sense, the philosophical and historical position of I. Kant is notable. According to the German philosopher, the beginning of history is shown as institution of human rationality (here, Kant actually interprets the biblical story of the Fall). Kant connected the final manifestation of human mind with a realisation of the exceptional position of humans in the nature: ‘the human understood that he is the purpose of nature, and the animals are the instruments. A sheep needs wool not for itself but for him (the human)’ (Kant 1786, p. 77). The natural expedience is here focused on expedience for the human. This anthropocentric position, according to the ‘recluse of Köningsberg’, shall be a prerequisite for accepting the categorical imperative, since every person, aspiring to be the purpose, must treat another person accordingly. In a rougher form, this ‘anthropo-teleological’ assumption is expressed in the famous phrase by I.S. Turgenev’s nihilist Bazarov: ‘the nature is not a temple but a workshop!’ (Turgenev 2015).

According to Bibikhin, the human, due to their artificial separation from the nature (forest), simultaneously separates from the experience of the perfect automat and therefore from Sophia of the world, plunging into the artificial world of technology instead. At the same time, the human can receive such an experience in religious ascetic and spiritual meditative practices, which are technical in the highest sense—*τέχνη τεχνών* (Bibikhin 2011, p. 32). In particular, the hesychast practice of ‘mental doing’ and continuous prayer. In this connection, it should be noted that the Russian philosopher understands the natural (forest) not as a thing lying (standing) opposite of the specific human principle but as a thing lying by (accepting) it.

We think that the Bibikhin’s phenomenological view of the Nature-Forest (Nature-Woods) described above required an amendment since any (even archaic) human experience is build upon the alienation-distance from the forest. We can find a good illustration of this thought in the novel *White Fang* by J. London. White Fang, a three quarters wolf and one quarter dog is by a twist of fate adopted with his mother into a Native American settlement. He is burdened by life there, the ‘voice of the Northern wilderness’ (i.e. nature-forest) calls for him and, in lucky circumstances, the grown puppy escapes. However, White Fang is already unprepared for living in the wild forest. We should note that it is not because he cannot adapt: the human settlement *i.e. home* (a place appropriated with human technology) has already entered him, left its mark on him and made him attached. A wild creature became domesticated, a wolf became a dog. The absence of signs of human presence (voices, smells and wigwams) burdened and depresses the puppy. As a result, the wild, ‘captured’ and escaped wolf voluntarily returns to the people, already domesticated, and, therefore, *their own* hound. A wolf in a sled harness shows both an archaic proximity to the forest and a distance (alienation) from it. The alienation from the Forest shows a way to the appropriation of a *Home*. The Home symbolically shows the technical-human principle in being.

Referring to an already mentioned metaphor of a fire in a forest by which sits the humanity, we should say that fire very clearly indicates a specific place in the forest. It is warm by it, but it may be cold in the forest; it is bright by it, but it is dark in the forest in the night; it is safe by it, but it is dangerous in the forest and so on. Co-being of the Home (technology) is an event of appropriation of a certain place in the world. The Home as a root and a lineage, as culture and technology is a result of attachment, anchoring and domestication. The ontology of the Home (ontology of

Technology) is ontology of *appropriation*. If hair growing on the body may be read as a manifestation of the forest (Bibikhin 2011, pp. 17–18), then clothes covering the body are therefore the home. Of course, we cannot speak about a simple binary 'mine-not mine' opposition. The hair does remain one's own but it is in a way alienated-distanced with clothes by covering and employing a utilitarian function (thermal insulation, protection). Clothes do not become a second skin in its direct meaning but actually perform the same function. It should be noted that in co-being of immersion into the forest the co-being of the home is present as well (in an open or hidden way). It can be explained by cases of adaptation to, 'integration' into the nature. For instance, 'subjective' accounts by experienced wildlife tourists about 'turning wild' in a sense of practically 'animal' adaptation to nature are very typical and common. Such adaptation occurs following sometime (10–20 days) in a relatively long expedition. A good example may be taken from diaries by N. N. Miklouho-Maclay: '…I'm turning into a Papuan: for example, today in the morning, feeling hungry when taking a walk and seeing a big crab, I caught it and ate it raw, i.e. live, everything I could eat in it' (entry of 7 March 1872) (Miklouho-Maclay 2011, p. 121).[2] However, even these extreme cases of 'integration' into the forest (nature) always happen to be connected with an experience of a home such as a tent, a place around a fire, a log cabin (as was the case of Miklouho-Maclay). Even a backpack worn by a seasoned forest traveller tells about the proximity and non-removability of the home phenomenon. It is also significant that many modern environmentally motivated decisions are based on an attempt to combine economic and ecological interests, for instance, raising the level of consumption without raising the level of energy consumption, increasing the share of renewables, etc. Economy is known to have etymologically originated from the ancient Greek οικονομος, i.e. home law, home care, home management.

It is crucial in the light of the aforesaid that the borders or the Home (Technology) and the World cannot fully match. Any declaration of such match is the intention of the New European subject who has considered itself the master of being. The Forest (Nature) in its disastrous manifestations can teach a lesson even to the secular New European mindset, showing it that the Home is by no means the whole World but a relatively (partially) enclosed and appropriated section (region) of the World. On the other hand, the Home (Technology) as a certain sum of the appropriated is subject to alienation. The alienation comes into the appropriated due to the fact that the appropriated is already as the appropriated. The Home (Technology) is in this sense encloses, separates its own thereby seemingly makes the possible mundane as the usual. The extreme of the appropriated world is its mundane nature which provokes another extreme: leaving the Home. What are the ways out of the extremes of the technological world (at Home)? Is the religious spiritual practice of approaching the perfect automat, and therefore the divine, described by Bibikhin, the only way out?

---

[2]Miklouho-Maclay, N. N. *Journey One* // Miklouho-Maclay, N. N. *Journey to the Maclay Coast*. Moscow, Eksmo, 2011, p. 121.

We think that there are several strategies it will take another study to classify and analyse. As one of the ways, we can view the practices where a certain 'meeting' of the forest and the house. The forest attracts people not only by the force of 'wine, tobacco, and drugs' but also by the dust of a road and the crackling of a campfire. The home, in its specific manifestations, (such as living premises, fireplace, clothes, heating, etc.) shelters the 'naked ape' from uncomfortable temperatures. Frost (cold) is often alienated from the usual situation of our lives, coming to us only as a negative background, difficulty and obstacle. A home for living becomes a place of enclosed warmth, with the cold being forced outside. In the same way, a modern urban dweller in situations of 'wind'' and 'cold' faces an alienating, 'unbridled' force of the North. Yet, there is an essentially different possibility.

The tradition of cold training is known from fairly ancient times. And quite early, cold was not used just for treatment, hygiene, and prevention (for instance, in the bathing practice of East Slavic tribes), but as a training method. We can speak here about the Spartan training system: agoge (Greek *αγωγη*—leading, bringing away and going away). The main goal was to train a male warrior in a boy. Personal development also involved cold training: clothing restrictions, sleeping on a cloth instead of a bed, long physical exercises in the open regardless of the weather, etc. Thus, apart from the classic ways of 'domesticating' the cold (clothes, heating and dwelling) by forcing it out in the forest, we can attest to a historical experience of appropriation of the cold. Here, we do not speak about separation enclosure from the forest but rather about strengthening the forest foundation, reintroduction of the human body into the forest. In modern times, a surge in mass interest in the phenomenon of winter swimming happened in the second half of the twentieth century. In particular, a number of city federations were formed and the first official mass events of the so-called 'walruses' took place in 1950s–70s in the USSR.

The very situation of immersion in icy water is a situation of partial exposure to the forest. Notably, the human being, in this case, turns naked or almost naked, i.e. deprived (or partially deprived) of the home-clothes. Of course, one may sceptically remark that water (lake, river, or sea). Yet this opposition applies to just one of the meaningful aspects: elements. It is significant that those who do extreme cold training and swimming are called 'walruses' in Russia, 'polar bears' in the USA, and 'seals' in Finland. We think that the practices of the 'extreme' appropriation of the cold require a detailed philosophical-anthropological analysis. The experience of existential phenomenological understanding of the Forest and the Home presented in this article can be a theoretical foundation for it.

3. We think that the specific character of being of humans has long been based *on a balance of the technical (Home) and natural (Forest)*. The human civilisation has been unfolding of the basis of this, actually borderline, position of the human being.

A technical might as encroaching of the Home on the Forest inevitably changes something essential in the being of humans. What can we say about the ways of solving the issues of human-nature relations? Probably, we should be sceptical to extreme positions of a radically negative attitude toward modern technology (like neo-Luddism) or, on the contrary, a techno-utopian worldview where all the issues are temporary and insignificant. It is unlikely that the solution is in the sociopolitical reshaping of the world. For example, we can imagine a world order which could eliminate, for instance, the threat of a global nuclear war while alternative technocratic threats would always be present. In particular, a good example can be found in a series of works by brothers Strugatsky, famous Soviet sci-fi writers, where science fiction has an obvious socio-philosophical agenda. We particularly focus on the Strugatsky novel series with the so-called 'Noon World.' The world of the 'Noon' is a hypothetical future of humanity in the twenty-second century where elements of communism and technocracy are mixed. In the framework of a unified communist technocracy which, according to the works, embraces the whole humankind, there is really no grounds for a nuclear war, but no one is immune from other risks, for instance, from global consequences of scientific experiments described in the novel Far Rainbow (Strugatsky and Strugatsky 2009). Its plot is based on describing the behaviour of people faced with a planetary catastrophe caused by physics experiments. As a whole, we can express a general doubt in specific measures and programmes aimed at taming negative consequences and risks of scientific advancement.

Even more dubious is a possibility of solving the above-mentioned issues of being and anthropology with sociopolitical measures. Consequently, the question remains open whether a human being will find a place in an increasingly technological world. And if yes, will this entity entirely integrated in the technological world be human? It is also unclear whether we as human beings are capable of a principled and meaningful revision of our attitude towards the Home and the Forest. Is it possible to have a certain historic achievement by deliberate transition to a simpler material and technical life? For example, is it the way N. A. Berdyaev thought of it (*New Middle Ages*)? Or should we, agreeing with M. Heidegger, admit that thinking cannot directly solve problems of the modern technocracy (yet can prepare us for new era of being, for appearance of God)? We think that it is possible to answer these questions *only by understanding and revisiting the very historical journey of the humankind*, which is a journey of a primarily borderline entity building a Home in an inalienable and inevitable presence of the Forest.

References

Bibikhin VV (2011) Les (hyle) [Forest (hyle)] Nauka, Saint Petersburg

Kant I (1786) Mutinafilicher Anfang der Menschengeschichte Russian edition (1994) Predpolagaemoe nachalo chelovecheskoj istorii (Conjectural Beginning of Human History) (trans: Shapiro A.I.), in Kant I. Sobranie sochinenij v 8 t. (Works. In 8 volumes) vol 8 Choro, Moscow, pp 72–88

Miklouho-Maclay NN (2011) Puteshestviya na bereg Maklaya (Journey to the Maclay Coast). Eksmo, Moscow

Press-vypusk 2418 02 oktyabrya 2013 Zaderzhanie aktivistov Greenpeace v Arktike: obshchestvennaya ocenka (Press release No. 2418. 02 October 2013 Detention of Greenpeace Arctic activists: public assessment) in Russian public of opinion research center: VTsIOM https://wciom.ru/index.php?id=236&uid=1096. Accessed 10 Jan 2018

Strugatsky AN, Strugatsky BN (2009) Dalyokaya raduga (Far Rainbow) in Strugatsky AN, Strugatsky BN Sobranie sochinenij v 16 t. (Works. In 16 volumes) Vol. 3 Moskow, AST, pp 119–244

Turgenev IS (2015) Ottsy i deti (Fathers and Sons). Azbuka-Klassika, Moscow

Beyond the Green Shift—Ecological Economics

Ove Jakobsen and Vivi M. L. Storsletten

Abstract Sharp contrasts exist between ecological economics and green economy. Green economy represents a serious effort to "green" the existing economy by using different instruments from the toolbox of market economics. Ecological economics argue that it is necessary to make changes at the systemic level, if we are to understand and solve serious problems such as climate change, the increasing gap between rich and poor, financial scandals, war, and terrorism. From the perspective of Imre Lakatos' (1922–1974) contributions to the philosophy of science, we argue that green economy, on the one hand, represents changes in the protective belt in order to defend mainstream economy's hard core, growth, competition, and strategic planning. On the other hand, ecological economics question the hard core and suggests a deeper change towards, development, cooperation, and partnership approach. To enlighten the tension between these two different approaches, we look at the work of Karl Mannheim (1883–1947) who distinguished between ideology and utopia as driving forces in societal development. The tension between green economy and ecological economics generates both energy and direction in the change processes.

Keywords Green economy · Ecological economics · Ideology and utopia
Scientific research programs

Introduction

Sharp contrasts exist between ecological economics and green economy. Green economy represents a serious effort to "green" the existing economy by using different instruments from the toolbox of market economics. Ecological economics argue that it is necessary to make changes at the systemic level, if we are to

O. Jakobsen (✉) · V. M. L. Storsletten
Nord University, Bodø, Norway
e-mail: ove.d.jakobsen@nord.no

V. M. L. Storsletten
e-mail: vivi.m.storsletten@nord.no

J. S. Methi et al. (eds.), *Borderology: Cross-disciplinary Insights from the Border Zone*, Springer Geography,
https://doi.org/10.1007/978-3-319-99392-8_13

understand and solve serious problems such as climate change, the increasing gap between rich and poor, financial scandals, war, and terrorism. Ecological economics maintains that by implementing green economy solutions, we remain trapped within the very same system that caused the problems. William Rees reasoned that green economy, rather than helping to solve the major environmental and social problems has, instead, contributed to hiding the real problems behind a veil of green words and concepts (Jakobsen 2017, p. 163). Ecological economics go far beyond the mere green shift to ask, and answer, critical systemic questions.

To delve deeper into the difference between green and ecological economics, we turn to Imre Lakatos' (1922–1974) contributions to the philosophy of science where he distinguished between a research program's hard core and the protective belt of flexible theories. From this perspective, we argue that green economy, on the one hand, represents changes in the protective belt in order to defend the mainstream economy's hard core. On the other hand, ecological economics question the hard core of mainstream economy. To enlighten the tension between these two very different approaches to socially and environmentally responsible economics, we look at the work of Karl Mannheim (1883–1947) who distinguished between ideology and utopia as driving forces in societal development. From this point of view, green economy makes changes in the protective belt of mainstream economy to save the hard core of the ideology, while ecological economics criticizes the hard core and puts forward utopian solutions.

To elaborate on these fundamental questions, we have structured the article as follows; first, we give a brief description of the main ideas in Imre Lakatos' theory of research programs as "hard core" and "protective belt". Second, we examine the tension between ideology and utopia based on Karl Mannheim's interpretation. Third, we connect green economy to the existing ideology and argue that green economy only represents an adaption of the protective belt, whereas ecological economics represent changes in the hard core of mainstream economy. Finally, we discuss some of the main principles in the hard core of ecological economics.

Research Programs

The Hungarian philosopher Imre Lakatos defined research program as a combination of hard-core theoretical assumptions that cannot be rejected or altered without falsifying the whole program, and a protective belt of theories that function as a safeguard to protect the hard-core assumptions. He attacked the idea that scientific knowledge can be regarded as a cumulatively growing system of statements and theories. The evolution of scientific knowledge is characterized by long periods where the hard-core ideas are protected by a belt of theories that are adapted to changes in the environmental conditions. When the challenges become too dramatic, critical questions are asked about the hard core and the evolutionary period is replaced by irregular revolutionary leaps where the fundamental presuppositions of scientific understanding are challenged and replaced.

Research programs provide a framework that is shared by those involved. When the hard core is threatened, then the theories in the protective belt are modified in order to save it. Changes in the protective belt are considered expendable and may be altered or abandoned when empirical anomalies require adaptations in order to protect the hard core. Based on the research program's hard core, research is conducted on the basis that some fundamental principles are simply not open to critical discussion. In this regard, the hard core is similar to Kuhn's concept of a paradigm.

Lakatos divided the levels within a research program into negative and positive heuristics. Negative heuristics point at methods and approaches to avoid, and positive heuristics refer to methods and approaches which are preferable. While the negative heuristic protects the hard core, the positive heuristic defines limitations for modification and changes to the protective belt in a general direction. Lakatos defined the main concepts in his theory in the following way:

> the negative heuristic specifies 'the hard core' of the programme which is 'irrefutable' by the methodological decision of its protagonists, the positive heuristic consists of a partially articulated set of suggestions or hints on how to change, develop the 'refutable variants' of the research programme, on how to modify, sophisticate the 'refutable' protective belt (Lakatos and Musgrave 1982, 135).

Not all changes to a research program's protective belt are equally productive or acceptable. Lakatos argued that changes in the protective belt should be evaluated not just by their ability to defend the hard core by explaining anomalies, but also by their ability to come up with new explanations and understanding. He classified adjustments as "degenerative" if they failed to accomplish anything more than the mere maintenance of the hard core. Lakatos' model allows a research program to be progressive even if it contains some troublesome anomalies. He argued, essentially, that it is necessary to continue with a research program even if it has problems as long as there are no better alternatives. In order to falsify a research program, there must be a better one available.

Ideology and Utopia

The Hungarian sociologist Karl Mannheim (1936) contrasted ideology and utopia on the basis of their social functions. Ideologies are directed toward the past and are important to create and preserve the identity, individually and collectively, in a society. He argued that ideology contains explanations and interpretations of the social reality which preserve the status quo. Utopia transforms the status quo by setting out an alternative to the existing society. Utopias are future-oriented and important if we are to define the direction of, and the energy for, change. The French philosopher Paul Ricoeur (1986) nuanced the differences between the two concepts by arguing that ideology is an attempt to legitimize the power structures in the existing society, while utopia represents an attempt to replace power by something else. "Ideology is the legitimization of present authority while utopia is the challenge to this authority" (Ricoeur 1986, p. xxi).

The tension between ideology and utopia could be described as a dialectic process with two poles, one referring to actuality and the other referring to potentiality. In reality, utopian and ideological elements are often interweaved. Without utopia, ideology becomes static, and without ideology utopia becomes an illusion. To illustrate the distinction between ideology and utopia, we can say that ideology is imagination as a picture which describes what exists, while utopia is imagination as a fiction, re-describing what exists. To understand the time we are living in, we must step outside the established ideologies and develop a utopian position from where we can observe it. Utopia provides a critique of the current ideology by providing alternative values and norms.

Ideologies relate mainly to dominant groups and serve to comfort the collective ego of these dominant groups. Ideologies reflect the general way in which groups determine what differentiates them from other groups and establishes an identity that satisfies the interests of the entire group. In addition, ideology legitimizes authority and it allows the authorities to convince the public that its regulations are necessary.

Utopias are naturally supported by minority groups and therefore more commonly by the lower strata of the society. At the first level, there is correlation between ideology as stability and at a second level, utopia is correlated to change towards the (im-) possible. From Ricoeur's (1986) perspective, there are two explanations why the ideological mentality assumes the impossibility of change. First, an ideology accepts the system of justification, explaining the non-congruence. Second, the non-congruence has been concealed by factors ranging from unconscious deception to conscious lie.

Ricoeur concludes by observing that the correlation between ideology and utopia forms a circle, a practical circle: "the two terms are themselves practical and not theoretical concepts" (Ricoeur 1986, p. xxii). This means that utopia in one period of time could be the ideology in a later period of time. In order to make the circle a spiral, it is of the greatest importance to imagine new utopias as ongoing processes. A spiral in the dialectic of ideology and utopia would occur if "the utopia of a given society challenged current ideologies at least to the point where they could be reflected on and compared to alternatives" (Steeves 2000, p. 226). "The death of utopia would be the death of society. A society without utopia would be dead because it would no longer have any project, any prospective goals" (Ricoeur 1986, p. xxi). This is in harmony with Whitehead's (1967) assertion that a society without adventure is in full decay and the conjunction of ideology and utopia typifies the social imagination. Imagination is constitutive of social reality itself.

Green Economy and Ecological Economics

In order to understand the creative tension between actuality (the dominating ideology) and potentiality (a realistic utopia), we find it relevant to start with a brief description of green economy and ecological economics in the context of Lakatos' theory of research programs. As a starting point, we accept that both perspectives are based upon a willingness to solve the environmental and social problems. Inspired by the distinction between ideology and utopia, we maintain that green economy has made changes in the protective belt of mainstream economics without questioning the hard core. Ecological economics is revising the hard core of mainstream economics in order to change the whole research program. In other words, green economy focuses on reducing negative symptoms to save the existing economic paradigm while ecological economics claim that the problems are caused by the hard core of existing economic paradigm for which an alternative has to be found.

Supporters of green economy accept that in a short-term perspective, increased growth has the highest priority in order to solve the current crises in economy, ecology, and society. The goal is to reduce the damaging effects of established economic theory and practice but to do it without making any hard-core changes. Because the methods do not depart from the established hard core (ideology) we can see that green economy strives to, and achieves, little more than maintaining business as usual. As long as the argumentations for the changes in the protective belt do not challenge the principles in the hard core, then mere greening of business is acceptable (Fig. 1).

Business ethics and CSR are introduced as means to reduce the most criticized consequences of aggressive competition which is to be retained and safeguarded as one of the main postulates in the dominating economy's hard core. Green economists often argue that environmental responsibility is closely coupled to green growth because we need more resources to handle the most serious economic and social challenges. They also argue that greening the economy is an efficient marketing tool to develop a reputation based on ecological and societal responsibility. By these means, green strategy is presented as an effective instrument to increase a company's profits and its competitive advantage. In a long-term perspective, they argue that growth should be as green as possible.

Ecological economics differs from green economy in that it asks critical and apposite questions about the hard core of the dominating economy and challenges current theory: first of all, growth is not part of the solution but is rather, right on the

Fig. 1 Greening the economy (change in protective belt)

Hard core	Protective belt
Growth	Green growth
Competition	Business ethics/CSR
Strategy	Green strategy

Hard core (main stream economy)	**Hard core** (ecological economics)
Growth (quantitative)	Development (qualitative)
Competition	Cooperation
Strategic planning	Partnership approach

Fig. 2 Ecological economics (change in hard core)

contrary, at the very core of the problem. Ecological economists such as Herman Daly and John B. Cobb Jr. (1994) argue that it is necessary to implement a de-growth economy and, in a long-term perspective, shift the focus away from quantitative growth towards qualitative development. Further, ecological economists argue that competition must be replaced by cooperation in order to handle the serious anomalies. Finally, if we are to find solutions to enhance the common good then strategic planning must make way for a partnership approach (Fig. 2).

So, ecological economics offers an interesting alternative to the hard core of the dominating economy and we can see for sure that ecological economics is not focused primarily on merely finding new answers to the old questions by making changes in the protective belt. Rather, ecological economics focuses critical questions on the hard core. By asking, and answering, these critical questions it is possible to uncover new perspectives and develop a more life-enhancing economic theory and practice. Ecological economics, fundamentally, asks how we can develop an economy and a society based on peaceful relations between humans and nature, between humans, and within the individual.

Ecological Economics as Utopia

In order to encourage change utopia has to challenge the existing ideology by, on the one hand, providing a critique of the current ideology and, on the other, providing alternative solutions. The term ecological economics refers to changes on different levels - ontology, system, practice and human consciousness. On the pure ontological level, it is a question of change from a mechanical to an organic worldview. One consequence of this change is that the status of economics changes and falls subordinate to ecology. To adapt the economy to the limits of ecosystems, organic knowledge and understanding are essential. The idea that natural science provides knowledge that gives man power over nature has to be replaced by an approach in which the goal is to learn from nature and develop a knowledge and a value system which teach us how we can best work with nature and fulfill human needs and improve our quality of life. The implication of this reasoning is the radical change which says economic activity is not an end in itself but a means to strengthen the life processes in nature and society. In any kind of social and

Fig. 3 Change from mechanic to organic worldview

ecological system economics has only one valid purpose and that, at its best and most basic, is to serve life processes (Fig. 3).

From an organic perspective, the global eco and social systems are comprised of closely interacting and interdependent subsystems based upon dissipative structures. The earth itself and all its living and nonliving components are interrelated, and the human being is a member of this integral community and must strive to find a proper role in it. Every system is connected to, and depends on, all the others in continuous evolving processes. Ecological economics, as a heterodox tradition, "accepts the transformative power of human agency with emergent properties arising from a dynamic interconnected process of multi-layered social interactions" (Spash 2012, p. 44).

Efforts to facilitate a sustainable future are hopelessly and inadequately covered by the existing scientific traditions of objective disciplinary approaches. To capture the interconnectedness of nature and society ecological economics is based on a transdisciplinary approach. Kenneth Boulding, one of the most influential contributors to ecological economics, once said that "the pursuit of any problem of economics draws me into some other science before I can catch it" (Kerman 1974, p. 6). He was looking for connections between different fields of knowledge, for the threads of theory that would tie together nature, society, and economy at the systemic, business, and individual levels. Today, the transdisciplinary approach is being rediscovered, unveiled, and put to use rapidly to meet the unprecedented challenges facing our troubled world. In its search to understand life as life, the transdisciplinary field of ecological economics examines the relationships between ecosystems, social systems, and economic systems in the broadest possible sense.

To meet the challenges of sustainability, and that means locally, nationally, and globally, we need an integrated knowledge based on transdisciplinary research. In transdisciplinary research, a disciplinary cross-fertilization makes the borders between the different sciences more transparent, (even to the point of disappearing) and practice and culture becomes integrated. Different kinds of systematic dialogs (world café, Socratic dialog, open space, and reflective dialog) are used as a means to connect people from all the different areas. This type of research also includes reflexive distancing that makes it possible to detect phenomena and contexts that have not previously been visible. By changing the focus of putting the individual business at the center of study and instead study the business as an integral part of a broader context, it is possible to achieve new insights into several areas.

Accepting that nature and society are not like lifeless and static machines (Nicolescu 2002; Capra and Luisi 2014), a transdisciplinary approach enables us to discover the resurrection of man as a subject of his own discourse. Furthermore, this line of research offers, according to Nicolescu (2002), not the end of history but rather the beginning of a new stage in human history. "Transdisciplinarity is nourished by disciplinary research; in turn, disciplinary research is clarified by transdisciplinary knowledge in a new and fertile way. In this sense, disciplinary and transdisciplinary research are not antagonistic but complementary" (Nicolescu 2002, p. 45).

Because ecological economics develops a theory and a practice that initiate constructive interplay between the surrounding cultural and natural conditions, the practical solutions will be different depending on time and place. Future solutions will not be exactly similar to solutions suggested today and solutions will differ according to geographical and cultural circumstances.

At the systemic level, the hard-core principle of the dominating economy, claiming that the market is made up of autonomous competing actors, is replaced by a view of the market as an integrated network of interdependent cooperating actors. Thus, the focus shifts away from objects toward relationships. Since the individual has to respect broad public values, a transition is required away from the egocentric economic man towards the "I-We" understanding. Although local production for local markets is the ideal, it is, of course, necessary to open up cooperation through and between international networks as well.

Ecological economics, inspired by natural growth curves, (increasing rapidly first then stabilizing), makes it possible to initiate continuous development in the quality of life without increasing the consumption of natural resources. The focus on qualitative development points to major changes in business; many companies and, indeed, whole industries will disappear, and new ones, far more in line with ecological principles and humanist values, will appear and take over.

At a practical level, strategic planning is replaced by a partnership approach founded on dialog and network-based cooperation. Free competition on the world market means that small producers in poor countries have always been and will always be the losers compared to powerful multinational corporations. Poor countries are forced to accept free trade in order to gain entry to the markets and the result is that the big companies gain increasingly larger proportions of the global markets and, for the system to work, poor countries have to receive aid. This results in a vicious cycle where the quality of life deteriorates both among those who have too little, and among those who have too high consumption. Today, the resources are distributed in such a way that the gap between rich and poor constantly increases.

Instead of focusing on top-down solutions, based on the initiative of politicians and business leaders, the focus within ecological economics is based on bottom-up initiatives. The change occurs when those who work and operate in the practical reality at the local level create solutions based on direct experience. It will always be, of course, necessary in top-down initiated rules and monitoring (for example, through national authorities) to have the United Nations and other global organizations.

It is of great importance, if we are to reach the common good as a result of cooperation based on mutual respect and trust, to make room for regular reflective meetings, and to create and empower forums for communicative action based on collaboration. Enterprises have a close connection to their local culture, and, by integrating in dynamic processes, culture becomes the source of inspiration and the connective glue. The network includes different sectors (for example, businesses, practitioners, and research and education institutions) and develops by integrating actors on several levels. This may serve as drivers for development of mutual principles for peaceful coexistence between the actors both locally and globally, where the actors are free to develop their own solutions based on their specific situational and cultural knowledge.

In general, we can say that committed cooperation between involved companies, interest groups, government agencies, and organizations, represents an innovative approach to many of the environmental and societal challenges we will face in the future. An important reason to establish forums for communicative and dialog-based interaction is to ensure that all affected actors are drawn into the preparation and implementation of concrete solutions. In this way, the actors are made responsible in joint efforts to realize the goals of sustainable development. A network of creative actors has better access to information when it comes to making oneself visible both locally and internationally. An advantage is that they promote their work and products, get inspiration to improve their work and products, and learn by sharing their experiences. Access to resources (both financial and material) from local, regional, and international sources, will further develop their ideas for activities and products and projects.

When the abstraction of "leaders hierarchy" is revealed and the structures are flattened by a fruitful combination of bottom-up and top-down processes, this will facilitate the development of nested networks of activities with the vital importance of integrating culture with economy and society and nature. In the perspective of the triple helix, circulation economics has life-enhancing connections to nature and culture which modifies and stimulates economic activity. Ecological economics requires that businesses have to include ecological and social values in their decision-making processes. Humans are part of the ecosystem and the ecosystem is a part of humans. By including social values in its thinking, business helps to create (optimal) conditions for the quality of life. As it is impossible to transform the different values into a monetary scale, policymakers should be able to handle the three different values simultaneously (Fig. 4).

Fig. 4 Ideology and utopia

	Ideology	Utopia
Ontology	Mechanistic	Organic
System	Competition	Cooperation
Practise	Strategy	Partnership approach
Individual	Economic man	Ecological man

On an individual level, ecological economics has significant implications for the definition of the economic actor. Instead of focusing solely on increasing profits and utility (the economic man), the economic actors put more weight and emphasis on the natural and social implications of production processes as well as products (the ecological man). A practical consequence is that market communication will have to include information about the working conditions for the workers in the entire production process and the extent to which the production process meets environmental requirements, requirements for animal welfare, the health implications for all involved, including the consumer. As regards ethics, a good and moral life, according to virtue ethics, is a life responsive to the demands of the world. Central concepts are good judgment, justice, courage, and self-control. To possess a virtue is to be a person with a given complex mind-set. "The most significant aspect of this mind-set is the wholehearted acceptance of a certain range of considerations as reasons for action" (Stanford Encyclopedia of Philosophy 2012). Virtue ethics focus on the moral person's character characterized by the ability to be aware of, to identify and to handle, and moral dilemmas in real-life situations.

Change Processes

In the preceding paragraphs, we have described and discussed ecological economics by contrasting it to green economy. We think this is essential because there are good reasons why most people, including scholars, have problems differing between green economy and ecological economics.

To illustrate and concretize the differences, we refer to the tension between ideology and utopia and the two components of Lakatos' research program. Green economy is anchored in the hard core of the dominating ideology. This means that changes in the protective belt are accepted as long as they do not disturb the hard core characterized by competition, growth, and strategy. Ecological economics represents utopian thinking, that is, a futuristic description of a potential society based on cooperation, development and a partnership approach. Within the vocabulary of Lakatos', ecological economics represents a scientific research program characterized by a hard core (negative heuristic) which includes fundamental assumptions that are completely different from the previous one.

When the hard-core principles change, the previous knowledge is re-interpreted and the whole research program changes; new questions are asked, new methods are introduced, and changes are implemented in practice. The change is like a Gestalt Switch where the context of understanding changes from economy to ecology. Instead of seeing nature merely as an economic resource base, our economy actually learns from nature. According to Colin Tudge (2016), all that stands between us and a glorious future is a bad strategy based on false ideas rooted in a debased ideology that puts short-term wealth and dominance above all else; in other words, the very definition of the hard core of the dominating economy.

References

Capra F, Luisi PL (2014) A systems view of life – a unifying vision. Cambridge University Press
Daly H, Cobb Jr JB (1994) For the common good – redirecting the economy toward community, the environment, and a sustainable future. Beacon Press, Boston
Jakobsen O (2017) Transformative ecological economics. Routledge
Kerman CE (1974) Creative tension – the life and thought of Kenneth Boulding. The University of Michigan Press
Lakatos I, Musgrave (1982) Criticism and the growth of knowledge. Cambridge university press
Mannheim K (1936) Ideology and utopia. A Harvest Book, New York
Nicolescu B (2002) Manifesto of Transdisciplinarity. State University of New York Press
Ricoeur P (1986) Lectures on ideology and utopia. Colombia University Press, New York
Spash CL (2012) Towards the integration of social, ecological and economic knowledge. In: Julien-Francois G, Rolf S (eds) Towards an integrated paradigm in heterodox economics. Palgrave Macmillan, Basingstoke, pp 26–46
Stanford Encyclopedia of Philosophy (2012)
Steeves JB (2000) Utopia and text: Ricoeur's critique of ideology, symposium IV, 2, pp 221–235
Tudge C (2016) Six steps back to the land. Green Books, Cambridge
Whitehead AN (1967) Adventure of ideas. Macmillan, New York

Myth on the Border Between Man and Nature

Irina Gorina

Abstract The article analyzes the peculiarities of mythological consciousness formed on the border of Nature and Man. The text deals with the process of the symbolization of the natural world in mythological consciousness is considered in which myth becomes a means of domesticating and humanizing the world, whereas the imbalances in the interaction between Man and Nature coincides with eschatological motifs.

Keywords Myth · Man · Nature · Existence · The process of symbolization Symbol · Consciousness · Border · Creative thinking

Diffusion of the Myth in the Borderlands of Nature and Man

Myth is the most important phenomenon in the history of culture and thought movements. In the early stages of society formation, the myth occupied a Central position in the process of developing the world perception and natural world cognizing in people. A characteristic feature of the mythological understanding of ourselves and nature is its diffuse character of mutual penetration and existence indivisibility of the individual with objects in the world (Tokarev and Meletisky 1991, 12). In these circumstances, the problem of determining the boundaries between man and nature cannot be precisely installed. Boundary, but rather a space of relationships being built, has provided direct experiences, the emotional affects, and spontaneous human actions. Thus, the myth did not fix the insurmountable boundary between nature and man, but was a kind of symbolic marking of real-life objects that people should pay attention to, developing optics and vigilance of his thinking.

Researchers S. A. Tokarev and E. M. Meletinsky indicates that the result "was naive humanization of the natural environment and a wide metaphorical" mapping

I. Gorina (✉)
St. Petersburg State University, Saint Petersburg, Russia
e-mail: gorina.sb@yandex.ru

J. S. Methi et al. (eds.), *Borderology: Cross-disciplinary Insights from the Border Zone*, Springer Geography,
https://doi.org/10.1007/978-3-319-99392-8_14

of natural and cultural (social) objects. People transferred their own properties to the natural objects, attributed to them life and human feelings" (Tokarev and Meletisky 1991, 12).

Metaphor, personification, creative thinking, and derivatives from sensual perception have formed in man a special area of imagination, in which the meeting with the objects of nature at the level of inner experience manifested an external act of this object's presentation in form of song, dance, picture, and legend. The aim of the myth lies in its public representation and joint, intense involvement of an event that comprises each individual who becomes a witness of visible first act of consciousness.

Whoever represents the myth is learning and knowing (shaman, mage, dedicated, or initiated into the unknown secret). The myth is the result and the line between knowledge and ignorance, familiar (i.e., filled with known signs) the place of stay or unfamiliar. Thus, the epistemological setting of the myth is the person's definition of his knowledge limits when they meet with objects of nature. In this case, it can be argued that the myth does not hold the boundary between nature and man, and is the limit of understanding its self.

In the myth creation, one can see the following universal positions. First position: this is the space where people face unfamiliar phenomenon or object. Second position: this is an introduction to the "host" or "hostess" this space, that is, with the spiritual substance of the phenomenon or object, which the Romans called "genius loci" or the genius of the place. Third position: establishment of contact with it when a man gives it the imaginative name, based on their internal sensory experiences at the moment of meeting. Fourth position: establishing contact with the natural object in the form of order relations in a situation of deceit, warfare, and games, the result of which is to receive the gift that anything can be traded. A gift is a thing or a certain, skill, and ability (such as mastery of fire, the invention of the boomerang or bow and arrow, the discovery of pottery, agriculture) becomes the property of transforming man and his existence.

Using mythic-creative positions, a person ceases to be an ordinary man. He becomes a "cultural hero" who performs a deed in the act of consciousness namely, it brings new knowledge to people—"new myth." Traditionally, the study of the mythology of the cultural hero is characterized, according to the theory of "historical poetics" of E. M. Meletinsky, by the following features:

> Cultural hero is primal forefather…. Strength, intelligence, cunning, magical and military skill are not differentiated in him. This complex character may be the Creator… and as a heroic Bogatyr who destroys monsters, disturbing the peaceful life of people… and as a comic trickster…. (Meletisky 1986, 22)

The main thing for a "cultural hero" is the actions and knowledge necessary for the survival of his family and tribe. In the context of our reflections, the concept of "culture hero" can be made wider. Indeed, a situation of sensory-affective experience and the connection to the ancestor was felt by everyone who played the myth about him and everyone who witnessed the performance. Cultural hero in the structure of the myth had an important function in preserving the unity of nature and man, organizing the formation of the cosmos as fixed in minds the order of all things in the known world. Thus, begins our insight into the structure of time as a dichotomy. Here, appears the distinction between mythological, sacred time in which ancestors live in empirical time of real life. Australian indigenous people described mythological times as "Dreamtime"—the time of the "eternal people" and totems (Meletisky 1991, 68). There are no boundaries between mythical and empirical time in space of external life. This boundary passes through a state of altered consciousness, namely sleep or performance of the myth in the ritual action. The diffusion of mythological and empirical time is obvious and related to the fact that time in primitive thinking is not characterized by duration of external processes and internal states, but by duration of stay in the rhythms existing binaries of the universe: life and death, day and night, and peace and war.

The myth is associated with an action. In Russian culture, there is a children's game in which boys and girls should share what they did and learned about the world. But the children's answer is: "We won't tell – we show." And, they are satisfied with pantomime, dance, or make hand gestures, which the audience must guess what the children did and learned that day. This children's game dates back to mythological thinking.

As pointed out by Levy-Bruhl L. In the monograph "Primitive thinking": "to speak with hands is to some extent literally think with hands…. Both languages are so different in their signs (one language consists of gestures, and the other from articulate sounds) are close to each other in their structure and in their way of expressing objects, actions, and condition." (Art Culture of the Primitive Society 1994, 68).

Two languages: speech and gesture have limitless possibilities of naming the world in the novelty of individual experience. In this connection, the myth of Adam, ancestor of all people, is interesting. When God created Adam from clay and put him into the Garden of Eden, he gave him three tasks. First: to cultivate a garden, that is, caring about his place. Second: to protect all the inhabitants of the garden and bear responsibility for all of them, and third—to give all objects of the Garden of Eden names. But the names were not predestinated because Adam was clean, not knowing the bitterness of the fruit of knowledge. Adam gave names to all objects of the world from their own observations of their existence and according to them. As a result, the name was a symbol, a sign, symbolizing the natural life creatures and a symbol in the mind of Adam. Such a notion of myth can be characterized as the result of an act of symbolization of individual objects and phenomena of the world in his mind.

Symbolization Process of the Nature World in Myth

Man as a bio-sociocultural entity is originally a product of organic nature, in need of its energetic resources of water and air, animal, and plant life. The primary instinct, self-preservation inborn programs of behavior, food procuring, and reproduction are inherent to him, like to all mammals. However, consciousness and mental activity, based on the thinking processes, create a specific model of relationship between man and nature.

Nature is pre-given to human thinking and perception is the main object of his senses: sight, hearing, touch, smell, and taste. In the process of ontogeny and phylogeny, nature can be considered as the primary sphere of human existence and the kind of the world territory. The world of nature is characterized as a reservoir of diverse organic objects, causing feelings of fear, aggression, and pleasure. The peculiarity of human perception and empathy in the world is the conscious, inclusive observation. At the level of consciousness and extreme activity of senses, a person turns on activity-thinking, in which there is a spontaneous decision in case of safety, self-defense, and fear overcoming.

Thinking is the secondary act after direct impression of objective reality, the derivative of which is the process of object symbolization in the consciousness. Symbolization of the object that caused the strong sensual excitement is the reverse understanding of oneself. For example, dangerous predators like lions, tigers, wolves, and bears were labeled in the individual consciousness with the title "my fear", elks, bison, deer, rabbits, and squirrels—with "my procurement - my luck", observing the fox, the snake opened qualitative characteristics of deceit, trickery, and lies. Thus, the position of "included observer" allowed to overcome the victim complex and through the determination of the qualitative characteristics of the "I" objects in the world - to understand and fix your own "I", that is, to find oneself through another, different on the level of consciousness.

Two vectors can be identified in this process: the first one—from nature to man and the second—from man to nature. Through thought man's consciousness became a perfect mirror of nature. Symbolization is dynamic in nature, depending on the will of man, his focus, emotional inclusion, and mood.

The symbol is a kind of retrogressive—retention traces of thoughts. It constitutes the surrounding nature objects in consciousness, creating a picture of the world in linguistic structures, sign systems. Retention traces of thoughts form memory space that holds the "I" definition of all subjects and objects in the world in the inner world of human. Determination of the subject's "I" is recorded in its name, in this sense, all nouns in the language on the mythological level of consciousness, are factors of hunting for the objects of the world. Hunting as a major real activity and conscious activity on the mythological level of consciousness has not only the exterior coverage of hunting lands, but also the inner landscape of awareness. Marking of the natural world in the mind takes place through the name of the objects. The object name is a symbol. It can be assumed that the original linguistic constructs formed as nouns.

As a result, on the level of awareness of the relationship between man and the caught "I", the rituals of animism, totemism, fetishism and shamanism are lined up. The goal is simple—to hold the "I" on the mental level in the format of early magic practices. Magic practice is reduced to the spell of the soul that is calling to one's desired object. The name is a perfect trap for the soul—the energy counterpart of the physical body. Calling on the spiritual level of the known object sooner or later leads to its bodily appearance. Through definition of "I" of objects, others, there is the predetermination of one's own "I" in space. The position of "inclusive observer" is overcome and a private place of existence is found.

The place of permanent presence and existence can be characterized as a "house of consciousness", where all "domesticated", "named" items are present. In fact, names become frontier posts to understand the world in the circle of life, thus, delineating one's own, familiar territory of rules for farming, fisheries, from the strange, the unknown world of nature. This is the area of the nearest world, the inner circle of the whole of all domesticated "I". According to Professor B. G. Sokolov, it can be characterized by the Greek term "oikonomika" (from which the modern notion of "economy" is derived) as the presence of "transformative action that changes the world" (Sokolov 2015, 41). Man, being present in the world as a repository, builds his house as the repository for all his close "I" of tamed nature objects that are necessary for his existence. On the level of human consciousness to tame means to humanize, make one's own.

Mythological Type of World Outlook

In the academic culturological dictionary "Zoonyms of Kildin Sami dialect", there is a story of a deer-werewolf Myandash—founder of the Sami, existing in the frontier of the world of nature and man, it acquires the appearance of man and beast. Myandash in the human form taught people the art of hunting, gave a bow, and prohibited killing female deer and leaders deer. Symbolization of the world of nature and man in his image has the ultimate nature of the universe existence: "In the Sami myth Ayeke-Tiermes hunted the golden-horn reindeer Myandash; when a deer gets the first arrow, mountains spew fire, the river will flow back, springs will run out of water. When the second arrows gets into Myandash's forehead, fire will cover the earth, people will begin to boil. When Tiermes's dogs catch deer and god plunges knife into his heart, the stars will fall from the sky, the sun will sink, the moon will extinguish, and only dust will remain on the earth" (Zoonym 2015, 86–87). The eschatological prediction contains syncretism of God's action as of a human being in the process of hunting and sacred sacrifices—the humanized deer as the whole organic world. In this and other symbols, a person captures one's own I-presence in the understanding of all things in the space of heaven and earth.

But if there is an "I" that is the opposite of "you", which determines the ratio of "one's own – another's", which is either evolving in the following positions "foreign - other – one's - own" moving to the simplest mental formula "you are me" or "you are mine", or remains hostile and unknown and unexplained. Even potentially dangerous objects and phenomena that threaten human, but explained by him lose their utmost cruelty, identifying with his organic being. Thus, the pronoun "I" in the aspect of symbolization of consciousness connects the two procedures: the detection of the object's existence and its immediate identification directly with a specific human presence in the world. The reflection of this process is a form of the myth—a myth—the story of the "I" of the object.

The time of development takes place in certain conditions. Conditions associate the object with the place. Time and space as the length measures do not exist and are unthinkable. The formula "here and now" is due to a period of existence as the detected result of the direct liveliness of perception and man's relationship with the object world.

In the myth, there is the presence of actions as a continuity of relations of man and nature. Relationships are built in a circle of existence, generating the Cyclism of time. The time cyclism is not tied to a gradual transition from one stage to the other but has a closed character of "beginning and end", which are not discernible in the magic act term. Terms of development takes place in certain conditions. Conditions associate the object with an object location. Time and space as an extent do not exist and are not conceived. Time-bound existence in the myth is the formula of "here and now" as a direct consequence of detectable vividness of perception and man's relationship with the object world.

Drama of magic action associated with the ritual practices of shamanism, initiation, and crafts shows the first presentation of the history of both the event and the forerunner of existential act. It was built not on the basis of appropriate activity, which has the end result of a change in quality and experience of the subject, but with the act of recognizing oneself in the represented object and becoming it. The name and the conditioned term of the object appears identical to the myth—a product of consciousness, which can also be characterized as a symbolization of the sensually perceived image in the manifested magic action. That is, the shaman in the mythological act becomes a raven, wolf, or the spirit, he overcomes the border of the closed form through the consciousness and becomes a symbol, positions oneself as a different object. The experience of mythological action allows to recognize the universal integrity of oneself and the world. In fact, a person identifying himself with the universe at the level of feelings, finds fullness of world perception as an early form of systematic knowledge of names defined by myth.

Myth is the border area in the human perception of the surrounding world. Structurally, the myth can be described in the format of presentation of the history of the world, or of any object through the direct experience of the magic of the song, the dance, and the rhythmic music. The mythological history of the subject and procedures constitutes human consciousness not in relation to external nature of the phenomenon, but to the spiritual component, hidden behind the outward form. Myth addresses the previous time, and is associated with memory as a recollection about the appearance of the object in the world. In this sense, the myth is always old, a construct of thought having appeared from the depths of one's memory. That is a phenomenon of the external world is just an excuse, the reason to call the myth, and a myth, as a product of thought, organizes semantic traps to hold events in the mind.

The myth is derived from the symbol, it is its external manifestation in the space of imagination, in the syncretism of sense and reason. The myth forms the "artistic image" through dance, word, music reproduction, and pantomime. Artistic living of the world at the level of the mythological worldview gives the mental image perception. The myth is a more complicated construct than the image, as the myth is the story of the genesis of the object formed by the thought and reflected in consciousness. However, the artistic image of the mythological worldview is a form of thought. Its foundation is the "sense perception". In this case, we can argue that on the level of magical practices, mythical thinking there is the syncretism "homo aestheticus" with "homo oiconomicus". "Homo aestheticus"—"a feeling person" is the definition of man as a creature marking his presence in the space of artistic images and art objects (Sokolov 2015, 10–11). The myth is a versatile machine of consciousness, which at the level of everyday existence creates the unity of the internal process of symbolization of objects in the world and the external process of organizing economic life, combining the simple logic of the action with its image. And this mythological perception of the world on the everyday level is a simple chain of causes, course, and consequences of vigorous activity, and on the image perception opens semantic integrity of the event. The Australian myth "Giving life" displays two simple actions of good spirits Leuna (Sun) and Calla (Rain). They relate to the emotional states of people: smiling Leuna, crying Calla and by uniting together, they give people the great many colored rainbow and "another invaluable wealth - the beauty" (The First Boomerang 1980, 10–11).

The myth can be considered as the first model of dialog between man and nature. In the dialogic model of the myth, a person does not take the position of a passive observer. He is included in direct experience. Nature appears as a collection of phenomena, whereas man, accumulating phenomena in the acts of consciousness, gets an idea. Presentation is a memory of objects that can be kept in the mind long enough, thanks to thinking. However, the boundary with nature is not observed and but experienced as a place of human constant meetings with its objects. The boundaries are everywhere. Moreover, each sign, superstition, life event, taboos, and prohibitions system are justified by the myth.

The myth is based on the typology of human relationships in society and in nature. World knowledge as a complex of representations about superstitions, ritual actions, events, and organizes human existence and the world of one's material and spiritual culture. Presentation, literally, is what is represented in the mind of man and the outside world, with no distinction in sense perception. That is, representing a bird, a man becomes a bird, representing the stone, one becomes a rock. Distinction at the level of sense perception between subject and object does not exist, it appears in the form of expression of language and linguistic elements of distinguishing the noun, the verb as a mode of action and the adjective as a qualitative determination of the phenomenon or object.

Myth and Language Structures

Language structures rationalize the myth, and in the perspective of the formation of historical forms of religious and philosophical outlook overcoming of mythological sensuous syncretism takes place on the level of consciousness, with the manifestation of dividing the world into this-worldly and otherworldly religion (i.e., the complexity of sensory perception in extrasensory phenomena of faith) and the priority of rational perception procedures in Philosophy, where one of the objectives was to clean up the language and consciousness from mythological ambiguity of artistic and imaginary representations of the world. In religion, the symbols in mythology—archetypes became either unused or defined in terms of relationship with God as a supersensuous personality, absolute of all forms of life. Philosophy interpreted them from the standpoint of abstract thinking in the context of a specific method of metaphorical speculation.

But in both cases the "I" of man has to be separated from the world of nature and abstracted from it. However, the mythological worldview type has been dominant for centuries and millennia, so the myth stands for any religious form and philosophical system, as a universal vehicle of culture, often restoring and adapting famous abstract and dogmatic metaphysical schemes to the primary layer of a simple and clear sensory perception of the categorical imperative as the starry sky or risen God as a sprouted grain in the ground.

In this regard, an interesting parallel between linguistic models of mythological thinking and the contents of the "Tractatus Logico-Philosophicus" by Ludwig Wittgenstein. It may seem paradoxical, but the text of the treatise can be described as the product of "sophisticated mythological creative thinking" (Ludwig 2005, 18). And the main topic of consideration is not Wittgenstein's logic of language, and what preceded, is the first beginning of its formation.

The main question for Wittgenstein is not the logic of language, but rather that which precedes it, that which is the first principle of its formation. The author of the treatise is concerned not with the question of thinking, but with that of the manifestation of thoughts. But "expressing thought in speech is like having a coat turned", it marks the boundary between the expressed sense of a judgment and that which remains unuttered in the zone of silence. Silence is, on the one hand, a consequence of the unrealizability of statements due to the impossibility of formulating the excessiveness and the intensity of internal experience; on the other hand, it is the lack of a situation which could translate silence into speech.

In the history of the myth, there is a similar problem of expression. In the words of V. B. Iordansky: "An important feature of collectively held archaic notions was that they contained abstract ideas that were expressed in a highly concrete form" (Art Culture of the Primitive Society 1994, 77).

According to Wittgenstein, the very ability of thought to capture that which is objectively present is made possible by the existence of "certain states of affairs" (Ludwig 2005, 23), that is of that which a human being can touch, feel, and otherwise immediately perceive. And, this is a peculiar manifestation of a particular sense-specific logic of the Myth, thanks to which, on Iordansky's view, "the archaic mind had developed a generalized and holistic model of the universe…the results of the abstracting activity … did not take the form of abstract theoretical constructs: they retained the concreteness of the imagery of the kind of reality the essence of which they revealed" (Art Culture of the Primitive Society 1994, 77).

The logic of the Myth shapes the world like an enormous "repository" of all things in which the human being has to determine causal relationships and which has to be converted into a cosmos. At the same time, the Myth was not a realm of illusions or delusions. Descriptions of the Myth reflect man's specific cultural and historical achievements in the social sphere (the use of fire, domestication of animals, and the invention of everyday objects and tools). It can be argued that in the Borderlands of Myth human beings think simply, clearly, and accurately, but their thought is based on the experience of their feelings and actions.

In his discussion of language, Wittgenstein analyzes the notion of a proposition. In a proposition, perceptible signs (whether sounds or written characters) are used as a projection of a possible situation (Ludwig 2005, 57). Signs, or symbols, that are used in specific, consistent, and simple propositional judgments, are correlates of Objects and States of Affairs (Ludwig 2005, 58). Thus, emerges a linguistic picture of the Myth together with its special esthetics of existence.

Myth: Esthetics of Existence

To paraphrase a saying of Descartes in the mythological consciousness the formula appears: "I feel, therefore I exist". This kind of position is characterized by a "dream time" of Australian Aborigines. Dream—as a way of sleeping consciousness was one of the first experiences of discovering the world of consciousness in signs and symbols. Unconscious living of one's "I" through what he saw, heard, felt. Way to oneself at the primary level of consciousness through overcoming fear, instinctive, animal states of aggression, fear, and pain. Birth of the primary text of the myth as sung, expressed in dance and pronounced in words state and relations to one's humanized world of nature (Moss 2017).

Mythological structures are the first stage of social development, the spiritual basis of self-organization and unification of people. For further development of historical types of religious and philosophical belief, mythological experience of symbolization in consciousness of the matter existent and manifestation of one's "I"—the presence of (homo oiconomicus) and "I"—presentation (homo aestheticus), becomes the defining vector of self-understanding in the world. Certainly, religion, and philosophy have their own specific features of symbolization of objects in the mind, thus revealing the complex process of becoming a person in existence.

The esthetics of modern man existence is not of sensory-specific or informative-technological nature. Informativity implies the transfer of the speech stream, endless communication. However, the sense disappears in the speech stream, which is not informative, but ontological. It is enclosed in the getting into the idea, but this is just coincidence. Following the judgment of Wittgenstein: "the World is everything that is coincidence" (Ludwig 2005, 18). In this case: "the world is the nature of coincidence", that is, things that happened spontaneously. This means that in the world there is no definitive finality, necessity, and certainty.

Randomness implies the existence of volitional effort of human freedom, since from the point of view of "Logical-philosophical treatise" in predictable the world disappears, dissolves into information that makes no sense. In myth-making the sense brings together nature and man, in internal presentation of their accidentally discovered relations. In today's world of global knowledge, of predictable actions and communications, technosphere replaces nature in the human mind. There appears the idea of Nature's secondariness in relation to Man and his technocratic progress. Technique is the tool for usage, but not an object of experience, any subject of the technosphere is made on standard and easy to be replaced. As a result, there is a new picture of the world already incorporating a third element: Nature-Man-Technosphere. In this paradigm, a Person is in the border of two incompatible without him levels of existence: the organic and the technological. The boundary between the worlds goes through the minds of every civilizational individuals that requires from them the will to live, which will set the balance between nature and technology. Myth: esthetics of existence.

References

Art Culture of the Primitive Society (1994) Reader. SPb, 16 pp

Ludwig W (2005) Tractatus Logico-Philosophicus. Selected works. M.: Publishing house "Territory of the future", 440 pp

Meletisky BM (1986) Introduction to the historical poetics of epic, Romance and Novel. M.: «Science», 320 pp

Meletisky BM (1991) Time mythical/myths of the world. Encyclopedia. M.: Soviet encyclopedia, vol 1, 671 pp

Moss R (2017) Way to time of dreams of early people. Electronic resource. https://psy.wikireading.ru/30877. Accessed 05 Feb 2017

Sokolov BG (2015) Ontology of sensitivity. (Conceptualization HomoAesteticus Part II). – SPb: Publishing House of RCHA, 292 pp

The First Boomerang (1980) Myths and legends of Australia. M.: «Science», 150 pp

Tokarev SA, Meletisky BM (1991) Mythology/myths of the world. Encyclopedia. M.: Soviet encyclopedia, vol 1, 671 pp

Zoonyms of Kildin Sami Language: Educational Culturological Dictionary (2015) Ivanishcheva ON, Mitina Yu S. MSPU, Murmansk, 108 pp

Part IV
Borders in Collision: Search of New Paradigms

Cultural Images of Nature as Paradigms of Human Practices

Aleksandr Sautkin

Abstract The article focuses on the role that cultural images of nature play in the human activity. The author shows that the Cartesian dualistic approach leads to the alienation of man from his own nature as a bodily experience. Imagination is considered as the link between corporeality and mind. Non-dualistic conception of human being is based on the structural theory of imagination of Gilbert Durand.

Keywords Man · Nature · Images of nature · Imagination · Bodily experience

Since a human being is a part of the nature, his attitude to it inherently includes his attitude to himself, i.e., the attitude to that natural element, which is contained in a human himself. Thus, with regard to the problem of "Human and Nature", we cannot consider only ecophilosophy, the issues of anthropogenic impact on the nature, prospects for the technological development of society, etc. One that defines our environmental practice and natural resource using are not in the area of environmental management (rephrasing Heidegger's passage on the essence of technique[1] (Heidegger 2000: 7)).

John Stuart Mill, a powerful mind of positivistic nineteenth century, gave the following definitions of the word "nature":

> we must recognize at least two principal meanings in the word Nature. In one sense, it means all the powers existing in either the outer or the inner world and everything which takes place by means of those powers. In another sense, it means, not everything which happens, but only what takes place without the agency, or without the voluntary and intentional agency, of man (Mill 1985: 375).

What exactly is the display of the nature origin in a human being, i.e., what happens in a human "without his own voluntary and intentional agency"? In our view, this nature origin is the body.

[1]"So ist denn auch das Wesen der Technik ganz und gar nichts Technisches."

A. Sautkin (✉)
Murmansk Arctic State University, Murmansk, Russia
e-mail: sautkin72@mail.ru

J. S. Methi et al. (eds.), *Borderology: Cross-disciplinary Insights from the Border Zone*, Springer Geography,
https://doi.org/10.1007/978-3-319-99392-8_15

We have to assume, the bodily experience is mediated by culture, language and consciousness (we use the knife and fork and abstain from breaking wind in the presence of others), but the human body physiology in its simple displays is not mediated by anything, and we experience it as something we cannot control: whether we have a headache, abdominal distension, or feel sexually aroused—all these do not depend on us and happen "naturally."

We deal with pain, but not with a thought of it. It is impossible to bring the corporeal experience into line with the experience of the mind. A thought is a mental representation of my pain or lust, but where is my pain or lust? In that connection, we would recall the discussion with the participation of J.-P. Sartre and G. Bataille, concerning whether it is the ego, who feels pain, the same one who thinks (i.e., Cartesian *Cogito*). For example, is the ego of suffering toreador, whose leg was pierced by bullhorn, the same one who thinks of it? Bataille believed that he is not.

Does a man have access to the natural aspect in himself?—Yes, he does.

Is it possible to make this experience direct property of the mind or to connect it to the soul in any different way?—At first glance, it seems to be impossible.

Here is a turn of Descartes' motif, which is so well known among the thinkers of modern philosophy:

> We can see that our soul, in so far as it is a substance distinct from body, is known to us solely from the fact it thinks, that is to say, understands, wills, imagines, remembers, and senses, because all these functions are kinds of thoughts. Also, since the other functions that are attributed to it, such as the movement of the heart and the arteries, the digestion of food in the stomach, and such like, which contain in themselves no thought, are only corporeal movements (Descartes 2004: 170).

In Descartes' writing "Passions of the Soul" thoughts, as the content of the soul, are divided into actions and passions (perceptions). Descartes considers imagination in two ways: thus, amongst "passions of the soul" he distinguishes those, which are caused by the soul itself, when we perceive ourselves as those who desire, i.e., perceive our desire as though from the outside: here he also speaks about "imaginations" which depend on will and constitute the actions (when we imagine a chimera, an enchanted palace or something which is only intelligible and not imaginable). On the other side, Descartes writes about passions which are caused by the body, they can be both dependent on the nerves and independent on them. Passions, which do not depend on the nerves, are also "imaginations" but different from those that were mentioned above, because from the Descartes' point of view the will has no part in forming them and they cannot be referred to the actions of the soul (they are "illusions of the dreams" or the daydreams). And since they do not depend on the nerves, they are only a shadow or a picture of the perceptions which the soul receives from the body through the nerves (Descartes 1989).

Thus, speaking about the body, we are trapped in a Cartesian situation: the body, i.e., the natural part in us, turn out to be alien to the soul as an element lying beyond the nature (in full accordance with Descartes I use the words "soul", "mind", and "consciousness" as synonyms, because eventually for Descartes *esprit* = *âme* = *raison*).

The body is what happens in us beyond our will and without our knowledge. This is what can be perceived by the soul and understood by the mind but as something different, alienated from myself; the content of the soul appears as something arbitrary and self-sufficient in relation to the corporality/the nature.

In 1937, initiating the first of his experience of the psychoanalysis of natural elements (just as a reminder, it was the "psychoanalysis of fire"), Gaston Bachelard aimed to rid the scientific mind and scientific notions from the influence of the corporeal intuitions which pervade the imagination. Later, he seemed to be quite enthralled (captivated and fascinated) with these natural elements—and now he considered them not as something that hinders the cognition, but as something that underlies all acts of a genuine creation.

In the introduction to "The Psychoanalysis of Fire", Bachelard writes:

> One would find beneath the theories, more or less readily accepted by scientists and philosophers, convictions that are often quite ingenuous. These unquestioned convictions are so many extraneous flashes that bedevil the proper illuminations that the mind must build up in any project of the discursive reason. Everyone should seek to destroy within himself these blindly accepted convictions. Everyone must learn to escape from the rigidity of the mental habits formed by contact with familiar experience. Everyone must destroy even more carefully than his phobias, his "*philias*", his complacent acceptance of first intuitions (Bachelard 1968: 6).

What is important for us in Bachelard's words provided above is the fact that he, science historian, acknowledges the undoubted influence of the imagination and unconscious corporeal (it is better to say "mythical") intuitions and representations on the constructs of our mind and in a broader perspective on the life of our consciousness.

Therefore, a human should avoid the dualistic thinking—he needs to find something in himself that links together the body/the nature and the soul as something lying beyond the nature (i.e., according to Mill, something that lies within the will and intentions of a human being).

In the first part of "Order Out of Chaos" (third chapter), I. Prigogine and I. Stengers write: "The nature must be described so that the very existence of a human will become clear. Otherwise, similarly to the mechanistic world view, scientific description finds its doubleganger in a human as in a machine endued with the soul and therefore is alien to the nature" (Prigogine and Stengers 1986: 131). No doubt the modern science, as Prigogine and Stengers also mention, exceeds the limits of classical concepts, overcoming their defects, but in the conversation about the human and the nature in many cases, we continue to be at the same level.

Actually, the point at issue is that in conversation about the practices, caused by rationality (also including technical practices), we should take into account the irrational and corporeal element of our being, i.e., speaking about relations between the human and the nature, we should take into account the nature in a human. From our point of view, this connecting link between them is the imagination, and the structural theory of the imagination, which was developed by French Sociologist Gilbert Durand and his followers, can be considered as a theoretical instrument of the research.

As Mavourneen M. Joy notes: "The image, in relation to this latter symbolic sense, has deep roots which reach below the surface meaning and find themselves grounded in fundamental, dynamic drives and configurations. These, Durand attests, have an intrinsic, spontaneous ability to express themselves in images of a radical nature that precede and underlie verbal articulation" (Joy 1981: 9). Any thought and any meaning eventually arise at the subconscious affective dynamic level.

The remarkable point in Durand's concept is that he seeks to implement a synthesis of humanities knowledge and natural science facts and builds his interpretation of the imagination, basing on the theory of reflexes developed by Vladimir Bekhterev and Alexei Ukhtomsky (and some other well-known Russian physiologists). From Durand's point of view, activity of imagination is determined by the operation of three systems of dominant corporeal reflexes which a newborn baby has: (1) a *postural reflex* (reflex of "standing" which expresses a striving for upright posture); (2) *digestive reflex* (reflex of feed intake); (3) *copulative reflex* (it is not only sexual reflex and even more likely connected with rhythmic movements (rhythm, repetition, yes/no, dance, and etc.)) (Durand 1992: 46–49).

According to Jean-Jacques Wunenburger, imagination is fixed in sensory motor schemes programming an "*anthropological traject*" (sic!), which comes from the neurobiological to spiritual and defines three types of representation constructions —they could be described as three verbs: to distinguish, to confuse, and to connect (French: *distinguer*, *confondre*, *relier)* (Wunenburger 2003: 52).

Imagination takes an intermediate place between ideal and material world. Imagination is a fundamental "in-between thing" but it is also a special world, not just a space and a border which doesn't have its own being.

As far as the world for us is a conjugation of the subjective and objective, imagination performs as a *primary reality* and is ontologically original because it is it that conjugates. This particular world is distinguished by Gilbert Durand as "*l'imaginaire*" ("imaginative" in its substantive meaning).

Therefore, figuring out our relation to Nature and analyzing corresponding practices one should focus on these archetypical models of *l'imaginaire* which breed a significant variety of nature images from century to century and from people to people; but they save some primary invariant schemes explaining to a human why at some moment of time he/she was oriented about nature in a certain way—in both senses of the word "nature," considered by J. S. Mill. (Figuring out symbolic structures of imagination could be found in the works of other philosophers, studying, for instance, the influence of religious beliefs on scientific knowledge development, independently of Durand's philosophy: Whitehead, Koyré, Needham and others.)

Now, there are some specific examples of today's relation of humans to a nature outside of them.

Example #1. Russian Federation.

1.1. (*negative*) According to mass media (September 2016), Russian Government considers the idea of total elimination of wild boars as a preventive step due to a threat of African swine fever virus spreading. Deputy Prime Minister of Russia

Arkadiy Dvorkovich claimed that "When we will be able to overcome this disaster, reestablishing boars' population will not be a problem, this could be done quite quickly." The executive chairman of National Meat Association Sergey Yushin said: "Wild boars should be seen less likely than elephants and leopards in the central part of Russia and other plagued neighboring regions." A number of expert zoologists protest against such measures (Lyutykh 2016).

1.2. (*positive*) The international agreement between International Crane Foundation, Directorate of Biological Recourses and Protected areas and Siberian Crane Foundation was signed in January 2016. The agreement is aimed on Siberian Crane (also known as Snow Crane) protection (Sakha Life 2016). That's just one of the measures aimed on saving this biological species: ornithologists in other regions of the Russian Federation are also working on this issue, particularly, growing these birds in captivity and then adapting them to wildlife. One can recall that even President of the Russian Federation Vladimir Putin took part in nestlings' fly training with the use of a motorized hand glider, in Yamalo-Nenets Autonomous Okrug in September 2012.

Example #2. Norway.
2.1. (*negative*)
When the International Whaling Commission put a moratorium on whaling in 1986, Norway and Iceland registered objections and continued to hunt whales commercially, as well as Japan, which hunts under a (controversial) scientific permit. Since 2014, when Japan briefly put a stop to whaling, Norway has become the global leader in whale hunting. Norway issued itself a quota for the hunting of about 880 whales for 2016. As far as few people in Norway want to eat whale meat, it is ending up in the feed manufactured for animals on fur farms. Jennifer Lonsdale, co-founder and director of the Environmental Investigation Agency, noted: "The Norwegian government claims it's important to have whale meat as a source of food for people, but because of falling demand, the product is now being exported. Now we discover it's going to feed animals in the fur industry, which we find completely unacceptable" (Bale 2016).

2.2. (*positive*) Norway is known for its extreme care for biological diversity and environment saving. There are many reserves and national parks (for example, Øvre Pasvik National Park which is closest to the Russian border). There is Atlantic salmon sperm bank in Norway, keeping samples of 6500 species taken from 169 populations. In 2013, as a part of the international program of rare wild animal species protection, a tigress from Novosibirsk Zoo (Russia) was taken to pair a young tiger living in Kristiansand Zoo in Norway (FlashSiberia 2015).

My commentary on these facts of humans' actions is quite short. At first sight, these facts contradict with each other, but they actually have the same root. Probably, human attitude to other species of our planet can be found out not by learning some demographic or consumer or any other material factors, but by understanding of who humans *imagine* they are.

I consider that such a treatment of human beings to other species is rooted in their excessive will, and as Friedrich Georg Jünger wrote, when a human loses the measure of his will, he becomes truly titanic being (Jünger 2006: 122). It nicely suites the image of "the Runaway of Gods" created by M. Heidegger; there, from where gods escaped, the titans, children of Gaia, arise. The Divine Cosmos is lost and is replaced now by Nature, understood as the mechanics of material elements, and the titanic human tries to control them, destroying and recreating the entire population, changing river channels and reshaping landscapes in the name of his own will, defining himself a measure.

Of course, this titanic archetype needs some further thoroughly analysis. "The titanic human claims himself only where life is considered as labor, and the world is considered as the world of labor. It appears in tremendous plans and efforts which excess any limit and fail pathetically when there no powers any more" (Jünger 2006: 122).

***

Let's go back to the theme of nature in human, in other words—to the question of the body and of the human attitude to himself.

As I said before, the body is something that is placed beyond our control and management. People take possession of their bodies in some degree but are not able to fully control it. Nietzsche's theory of *Übermensch* probably emerged exactly from his illness. A pain is not a thought, pain is something nonreflective. The thought of the pain comes to us post factum, and we are dealing with it as though with the pain—that is the result of our consciousness' working, that alienates us from our "nature," i.e., our body.

Arne Johan Vetlesen in his excellent book "A Philosophy of Pain" writes: "In a fundamental sense, the feeling of pain is not a product of thought. <…> the rule is that pain comes and goes independently of human though and will. The reality of pain is the reality of the feeling of pain. <…> Pain heaves the ego down from the pedestal" (Vetlesen 2009: 51–52).

The phenomenology of pain, so convincingly presented in the fifth chapter of his book, shows the immanentisation of *my* being, which is closed in the body and in its pain (i.e., in my pain, because I do not exist out of my body). However, Vetlesen also shows that the attitudes to pain, or "dealing with pain" are mediated by culture, and, in his view, modern Western European culture (as well as the Norwegian culture as its integral part) is based on the idea that our living is a permanent process of self-creation. He states that this idea has become a kind of ideology.

Epistemological a priori, preforming the human attitude towards one's own body as an element of natural order, is rooted in the imagination. In present days, we are watching the manifestation of the same "titanic" archetype in the techniques of dealing with the body. The idea of the controlled metamorphosis becomes a key idea, now it determines the value system.

Medicine is one of those "body technologies," the others are hygienics, body care, sexual discipline, physical training and sports, etc. We believe that imagination sets those frames in which the theoretical knowledge and practical technologies are developing. Medicine of Antiquity, medicine of Middle Ages, and medicine of the Renaissance were quite different, despite the fact that they had some common ground (Galen's ideas): people of different ages imagined their own body in different ways. It is obvious that previous historical eras could not have had the idea of voluntary intervention into the bodily processes to manage them. There is enough to remind the aphorism ascribed to Hippocrates: *Medicus curat, natura sanat*. Any "body technology" is a human intervention into the flowing of natural processes, but only following the Nature (*φύσις*) it can be successful.[2] The man of Antiquity, even in the fight against disease and pain, did not pretend to change the nature, but wanted only to follow it. Perhaps, this attitude explains one of the points of the Hippocratic Oath, namely the principal rejection of euthanasia and abortion. The ancient Greek physician denies to believe that the origin of life and its end may become an object of his intervention.

The modern situation requires the ability to interfere with the bodily processes. If I have a headache, I take a pill. If I have a toothache, the dentist treats it. Everyday character of these procedures obscures the fact of intervention of science and technology into the body for the task of its transformation. Besides, people are not satisfied with just getting rid of bodily pain as such if he/she isn't satisfied with the shape of his/her nose or lips, he/she may ask a doctor to change them. (What would Galen say about plastic surgery? We can assume that his reaction would be negative.) We can also see the intensive intervention of mechanical devices into our body: dentures and artificial pacemakers are just details. Somewhere on the horizon, there is an image of the *cyborg* as one of the most characteristic figures of the post-humanist thought. It is even possible to speak not only about the compensatory mechanisms, but also about the transformative intervention in the bodily processes at the molecular level: "nano-colonization" of human body.[3] We see an increasing intrusion of medicine into the human world. Now medicalization catches both bodily pain and "psychic" pain, such as suffering from depression and other negative psychological states. Pharmacology is also a technical phenomenon. Antidepressants as part of contemporary everyday life are an element of the same policy of corporeal nature's management. (Let us remember the movie "Prozac Nation" by Norwegian film director Erik Skjoldbjærg.)

Ivan Illich wrote in his book about medicalization of our living:

> To enable individuals to transform bodily pain into a personal experience, any culture provides at least four interrelated subprograms: words, drugs, myths, and models. <…> The cultural setting not only provides the grammar and technique, the myths and examples used in its characteristic 'craft of suffering well,' but also the instructions on how to integrate this repertoire. The medicalization of pain, on the other hand, has fostered a hypertrophy of just one of these modes – management by technique – and reinforced the decay of the others. Above all, it has rendered either incomprehensible or shocking the idea that skill in the art of suffering might be the most effective and universally acceptable way of dealing with pain.

---

[2]It is noteworthy that Galen puts medicine (ἰατρική) in the first rank of the arts (τέχνες) with rhetoric, music, geometry, arithmetic, astronomy, grammar and the law, as well as sculpture and painting, not considering it as something strictly applied like the art of a carpenter or shoemaker. See: (Galen 2014).

[3]The authors of science fiction stories often used this plot, describing the invasion of microscopic doctors inside the human body: for example, "Fantastic Voyage" (1966) by Isaac Asimov or "Swallow the Surgeon" (1970) by Soviet writer Georgiy Gurevich.

> Medicalization deprives any culture of the integration of its program for dealing with pain (Illich 1976).

We do not want to deny the importance of the fight against pain and the role of medicine in this fight. On the contrary, in many countries, the problem has not yet been solved (e.g., the lack of strong painkillers for cancer patients). Let us not forget that "pain heaves the ego down from the pedestal," pain deprives human beings of their dignity. However, we are talking about societal tendencies of Western civilization, and they are increasingly alienating the human from his own body experience, and, therefore, from nature. In extremum, these trends give rise to a post-humanist idea of liberation from human nature, which is embodied in our bodies.

References

Bachelard G (1968) The psychoanalysis of fire. Trans by Ross ACM. Beacon Press, Boston

Bale R (2016) Norway's whaling program just got even more controversial. National geographic official site. http://news.nationalgeographic.com/2016/03/160331-norway-minke-whaling-fur-farms/. Accessed 21 Dec 2016

Descartes R (2004) The description of the human body. In: Stephen G (ed) The world and other writings. University Press, Cambridge

Descartes R (1989) Strasti dushi [Passion of the Soul]. In: Sochineniya v dvukh tomakh [Writings in two volumes], vol 1. Mysl, Moscow, pp 481–572

Durand G (1992) Les structures anthropologiques de l'imaginaire. Dunod, Paris

FlashSiberia (2015) Tigritsa Bonya, pereekhavshaya iz Novosibirska v Norvegiyu, rodila chetvernyu [Arrived at Norway from Novosibirsk tigress Bonya gave birth to four]. http://flashsiberia.com/news/tigrica-bonya-pereehavshaya-iz-novosibirska-v-norvegiyu-rodila-chetvernyu. Accessed 05 Jan 2017

Galen (2014) O pobuzhdenii k meditsine [Exhortation to the study of the arts especially medicine]. Old Greek text and Russian trans by Balalykin DA, Shcheglov AP, Shok NP, in Istoriya Meditsiny [History of Medicine], 2, 141–160

Heidegger M (2000) Die Frage nach der Technik. In: Heidegger M (ed) Gesamtausgabe. Bd. 7: Vorträge und Aufsätze. Vittorio Klostermann, Frankfurt am Main, pp 7–36

Illich I (1976) Medical nemesis. The expropriation of health. Pantheon Books, New York. http://soilandhealth.org/wp-content/uploads/0303critic/030313illich/Frame.Illich.Ch3.html. Accessed 05 Jan 2017

Joy MM (1981) Towards a philosophy of imagination: a study of Gilbert Durand and Paul Ricœur. McGill University, Montreal

Jünger FG (2006) Grecheskie mify [The Greek Myths]. Trans by Shurbelev AP. Vladimir Dal', Saint Petersburg

Lyutykh S (2016) Operatsiya "Kaban". Pravitelstvo planiruet pogolovnoye unichtozhenie dikikh sviney v Tsentralnoy Rossii ["Boar" operation. The government plans to completely destroy wild pigs in Central Russia]. https://lenta.ru/articles/2016/09/23/kaban/. Accessed 05 Jan 2017

Mill JS (1985) Nature. In: Robson JM (ed) Collected works of john stuart mill, vol X. University of Toronto Press, Toronto; Routledge and Kegan Paul, London

Prigogine I, Stengers I (1986) Poryadok iz khaosa [Order Out of Chaos]. Trans by Danilov Yu A. Progress, Moscow

Sakha Life. (2016). Podpisano mezhdunarodnoye soglasheniye o sokhranenii yakutskogo sterkha [An international agreement on the preservation of the Yakut Siberian Crane was signed: press release of the Ministry of Nature Protection of the Republik of Sakha (Yakutia)]. http://sakhalife.ru/podpisano-mezhdunarodnoe-soglashenie-o-sohranenii-yakutskogo-sterha/. Accessed 04 Jan 2017

Vetlesen AJ (2009) A philosophy of pain. Trans by Irons J. Reaktion Books, London

Wunenburger J-J (2003) L'imaginaire. P.U.F, Paris

Imagination as a Breaker of the Border

Andrey Vinogradov

Abstract The topic stated in the title can cause confusion. Indeed, the notion of the breaker of the borders contains a clear negative connotation and when it is used as a characteristic of imagination, an impression appears that it is accused of something: whoever breaks the border violates the law. He becomes a criminal and should be judged for it. Let's try to understand whether imagination is guilty of anything, whose borders it violates, and whether it is necessary to punish it for that?

First of all, let us define the notion of imagination. In the broadest sense, imagination is understood as a person's ability to recreate in his mind the surrounding reality in the form of images. Such imagination never violates the borders as it is a reproductive work and there is nothing to accuse it of. Only creative imagination can be a breaker of borders because it is able to create something that did not exist, and thus breaks the usual stereotypes, changing reality. But even creative imagination cannot always be charged with violation of the borders, only under certain conditions this may happen. To understand this, you must use two philosophical categories: "immanent" and "transcendental". Under the immanent (from Lat. immanens—"staying inside"), we understand something which is inherent to the nature of the phenomena with which we are dealing, in both material and mental reality, and that has no basis in something external to it. And under the transcendent (from Lat. transcendens—"beyond"), we take something that has a different nature, external to the phenomena, to explain which you have to go beyond the border of the immanent.

Imagination (even creative) may not go beyond the immanent reality. If imagination creates something new by combining elements of the immanent reality, using the immanent logic, the inherent logic, it never invades, and violates anything. The appearing picture, perhaps, is not a real fact, but it is quite probable. It could arise in the case of a combination of circumstances or it is potentially possible in the future. But if the novelty is the result of appearance of elements that do not

A. Vinogradov (✉)
Murmansk Arctic State University, Murmansk, Russia
e-mail: andvinogradov00@mail.ru

J. S. Methi et al. (eds.), *Borderology: Cross-disciplinary Insights from the Border Zone*, Springer Geography,
https://doi.org/10.1007/978-3-319-99392-8_16

belong to the immanent, or it is the result of the logic that cannot be deduced from the immanent laws, in this case, we really deal with a “breaker”, whose guilt can be regarded as proven.

In view of the brevity of the article, only three aspects of this vast topic will be considered. First, we will discuss the methodology of consideration of the issue, second, we will study the specific manifestations of the imagination, and third, we will consider the effects that arise as a result.

With regard to the first question, it should be noted that its decision is dictated by the specificity of the subject reviewed. If it goes about exploring imagination in the first, the most general sense, then we could safely use the instruments of the rational methodology. Such tools are adequate to the intrinsic reality, therefore, they are quite suitable for this kind of analysis of imagination, which does not breakers its borders. But when the rational methodology is viewed as the universal key to all the locks, serious problems appear, not only in the perception of reality but also in the course of social processes. Many social upheavals were caused by the fact that the views about society and prospects were limited to the rational framework. The thinkers of the past (particularly during the preceding three centuries) thought that they had understood the rational laws of social development and, because of this, it became possible to make social changes.

One of the first prominent examples of such kind is the activity of French philosophers of the Enlightment period. Their teachings were characterized by a rejection of transcendental views. Court of reason was the highest authority in the struggle against obscurantism, engendered by religious imagination. The Enlightment philosophers for over a century tried to make people believe in the limitless possibilities of the human mind. But, imagination, based solely on the mind, generates a person’s dangerous illusions with regard to his capabilities. The scale for comparison disappears and the man begins to overestimate his strength. In France of the eighteenth century, this reassessment ended in the Jacobin dictatorship. So, errors in the methodology of cognition in practice may lead to the loss of many lives. Immanent imagination spawned illusion, not less dangerous than transcendent.

In this light what was written at the beginning of this article takes the opposite meaning. It turns out that innocent, at first glance, imagination, functioning within the limits of immanent reality turns out to be guilty of not going beyond these limits and, thus, clamps the human being in the immanent necessity. To avoid this, you should consider the subject much broader, including the concept of imagination and its transcendental manifestations. This, in turn, will require an adequate methodology, which is not limited exclusively in the rational sphere.

The history of philosophy shows that the ground on which the methodology of cognition of the transcendent was tested, was the theme of morals. The French Revolution (like other revolutions of the nineteenth and twentieth centuries) was, among other things, an empirical test of the hypothesis of the possibility of establishing morality based solely on principles, immanent to human nature, without any transcendental assumptions. People thought that they could create a morality as a rationally built construction and they have been still paying for it.

It is interesting that in the same eighteenth century, there lived a thinker who sought the complexity of the described methodological problem and found the way out. It is Immanuel Kant. On the one hand, largely echoing the spirit of his era, he sought to stay within the frame of the rationalist methodology. On the other hand, he saw the danger of forgetting the transcendent and thought much over its methodology of cognition and, I would say, its rehabilitation. He said that a higher entity that is not available for us in its transcendent state, can be postulated by moral law existing in us (Kant 1994, p. 381).

But this was the limit beyond which Kant did not want to go. Otherwise, he would have to start speaking about the properties of the transcendent, and that he considered unacceptable since the transcendent is not the subject of knowledge. According to Kant, the belief in the existence of God and the afterlife is a question of "not logical, but moral credibility" (Kant 1994, p. 484). So, it turns out that it is not the transcendental essence that provides the justification for moral claims, but on the contrary, our morality provides us with confidence in the existence of the transcendent entity.

Of this idea, in my opinion, the important methodological principle of the study of transcendent consists. This methodology involves, first, the mediated learning of the transcendent through its manifestation in some immanent phenomena. This is obvious because the transcendent is beyond any possible experience, so we can study it only through its indirect manifestation. Second, it is necessary to take for analysis those immanent phenomena, the nature of which cannot be separated from the immanent reality and to explain which we have to assume the existence of something outside its borders. The reality is that if we do not recognize the transcendent source of these phenomena, we generally are not able to explain their nature. This is exactly how morality, disconnected from its transcendent basis loses all meaning.

The biggest representative of the Russian religious philosophy V. Solovyov, who lived in the second half of the nineteenth century, was as well as Kant very interested in moral values and their relationship with the transcendent. However, according to Solovyov, Kant's methodology was characterized by subjectivity. In his book "Justification of the Good" (1899), Solovyev wrote that the author of "Critique of Pure Reason" destroyed the naive realism of his predecessors, and showed that our cognitive capabilities are limited to the phenomenal world. Indeed, if only phenomena are available for human cognition, then we will never be able to learn the essence of what we are dealing with.

Based on this provision, V. Solovyov finds fault with the validity of Kant's moral philosophy. Kant states that conscience obliges us to treat everything properly. Solovyov asks: If conscience, as a manifestation of morality, is only a mental phenomenon, then what is its binding force based on? And if it is something more than a phenomenon, then "the moral law has its basis not only in us, but also regardless of us; in other words, the absolute law implies an absolute legislator" (Solovyov 1990, p. 243). V. Solovyov brought Kant's idea of the transcendent source of morality, without which morality loses its objective and binding character, to a logical conclusion.

From V. Solovyov's point of view, open and honest recognition of the transcendent source of morality gives her a solid foundation. Kant's transcendental appears as an assumption necessary for the existence of the moral law. Solovyov shows the unreliability of this position: "to justify a pure morality the genesis of God and immortal soul is needed, but the accuracy of these ideas could not be proved, therefore, pure morality determined by them remains the assumption deprived of credibility" (Solovyov 1990, p. 244).

In order to get out of this difficulty, Solovyov suggests that the methodology proposed by Kant should be modified. He believes that the existence of the transcendent must be approved initially as a prerequisite for any human activity, rather than as a conclusion resulting from this activity. In his opinion, the nonrecognition of the transcendent makes all human life meaningless, so, admitting the transcendent must precede the consideration of any human activity.

Thus, to the two Kant's principles of the cognitive methodology of the transcendental, we can add the third principle, suggested by Solovyov, who claims the ontological status of the transcendent. From this perspective, imagination (even creative) does not create a totally new product but serves as a way to express a hidden content of the existing transcendent reality. This does not mean that imagination does not create anything new. It only means that the creativity of imagination is in the originality of the form by which it interprets the content of the transcendent reality. Imagination makes this content available to us. And the work of imagination should go beyond the boundary of the immanent reality, as it tells us about something which does not belong there.

Thus, we have come to the second question of the specific manifestations of imagination as the breakers of boundaries of the immanent world and we will try to apply this methodology to identify these manifestations. What does this methodology give for the analysis of morality? In short, it allows us to find a common basis for all moral views. If we consider the morality as the product of inherent reality, then it is a combination of contradictory provisions. Each nation creates its own ethical standards, which are not similar to other people's standards. But, from the standpoint of the methodology described above, their diversity is the only variety of forms, each of which in its own way expresses the same immanent content.

What are these phenomena, in addition to morality? First, all of them are related to the man, to his activities. The man is a creature that violates the borders of the immanent world. Only he can hack "the window into the transcendent". Second, the essence of these phenomena is that man begets them, committing acts not in conformity with the laws of the immanent reality, but in line with fundamentally different logic, fundamentally different motives, values, or goals. The transcendental invades the world through the human in nonreflexive and non-theoretic forms of his spiritual experience, in the areas where rational action turns impotent.

Where should we look for these violators of the borders of immanent? The most ancient sphere is mythology. Not all of it. There are elements in mythology that are generated by intellectual activities, elements that are sufficiently consistent with common sense, which means that they have an immanent origin. But there are moments that cannot be explained from the rational perspective, moments of profound insight, about the origin of which there are the most improbable hypotheses such as transferring them from aliens. Otherwise, it is very difficult to explain the presence of such astounding knowledge in ancient people that could not be achieved by using the methods of science of that time.

Common to all people transcendental reality breaks into our world through the diversity of myths. This commonality of mythological basis was very well shown by a well-known researcher of mythology D. Fraser. Detracting from the diversity of forms of expression used in the myths, he found a remarkable repetition of their plots. All the peoples of the world, not related, told similar stories about the creation of the world and the man, about the loss of human immortality, about the fate of the human soul, etc. So, the imagination helped man to present processes which are beyond the capacity of his perception through clear images. The diversity here is a fee for the complexity of the task, which is to express the ineffable.

The next border violator is religion. Also not in all aspects. It, surprisingly, has quite a lot of immanent. For example, theology is a completely rational activity based on formal logic and it uses a number of methods specific to science. Transcendent in religion erupts in those moments when people are removed from the habits and representations that are formed by the immanent reality and plunge into the perception of the world. Those can see it who has reached enlightenment, abandoning the goods of this world, in meditation practice, or prayer ecstatic. They see great mysterious eyes on the faces of the Fayum portraits and ancient icons. When the official religion tries to formalize it, legitimize it in certain phrases and rituals, the transcendent disappears, becomes immanent, it loses its spirit, and its life.

Religious imagination has created a variety of images. In this area, the diversity of forms is even bigger than in the realm of mythology. In our view, this is so due to the bigger complexity of the task: here we should not just talk about phenomena, which are sensory inaccessible to humans, but it is necessary to establish a relationship with them. As a result, the national customs and traditions that regulate communication, cultural peculiarities, artistic tastes, and other specific factors are involved in this process. But behind all this religious diversity stands one transcendent reality that compels a person to look for something more than the inherent givens. And imagination helps to embody this vague motivation into specific images available for perception.

Hand in hand with religion in violation of the borders of the immanent always went art. For example, you can recall D. Bernini's sculpture "Ecstasy of St. Theresa". Religious enlightenment and enlightenment of the artistic genius merged in it at the same time. Creative inspiration is a mystery, unexplained from the position of the immanent. In general, N. A. Berdyaev showed the analysis of the manifestations of the transcendent in creativity in great detail. According to him, "an act of artistic freedom is transcendent to the world's given" (Berdyaev 1989, p. 369).

This is the area where imagination reaches the highest magnitude. Art gives a wide space for creativity. The question is how to evaluate its results, what constitutes a work of art. Is this a copy of reality, or embellishment, or parody on it, etc.? Most likely, all the three options are right as well as many others may be. If we go on enumerating the variants, we will lose the essence of the phenomenon. Much more important is the question, what led the author when he was creating his artwork. Only those artworks find a response in the hearts of other people that are created by real inspiration, which, as mentioned above, is a mystery.

But these all are areas in which we are fundamentally ready to meet with something otherworldly. The most surprising thing is that we can detect the manifestation of transcendent in some unexpected areas. For example, in scientific activity. On the one hand, science by its essence seeks to explore the immanent world. It must reveal its immanent, inherent patterns. But some moments in the way it does so maybe transcendent. All, who are engaged in scientific activity, know the conditions, in which the intuitive insights occur. Such outbreaks usually appear on the background of fatigue from prolonged rational activity and they are not understandable from the point of view of logic of the studied reality.

Researchers of this problem note that rationality, which immanentism entirely relies on, puts science into too narrow frames, not allowing it to thrive. A modern solution of this question involves access to a much wider range of tools and techniques of cognitive activity, including artistic imagination which is not very scrupulous in respecting the boundaries of immanent: incomplete knowledge of the world can to some extent be compensated by the diversity of additional languages of descriptions included in the learning process, one of which is art.

Even in the most fundamental scientific concepts aimed to describe immanent reality, the traces of transcendence can be traced. It is enough to wonder what a scientific fact is. As it is known, the representatives of all variants of positivism tried to answer this question. Their desire was to represent the fact in the purest form, free from any metaphysics. But they could not do it fully. As always being only interpretation of reality, any fact contains elements of the unknown, and perhaps unknowable. In the light of this, the widespread view that science relies on metaphysics as its limit looks quite sensible now. Thus, also here transcendent invades the province of the immanent, manifesting itself as the breaker of the border.

In this regard, a natural question arises: what can be the result of such disorders. What can they give a person? First of all, it should be noted that manifestations of the transcendent look as the violators of borders only in terms of immanent reality. I would generalize that the penetration of the transcendent into immanent gives a person a chance. It invades the totalitarian Kingdom of immanent and, if it does not change it, then at least it illuminates it by his dream of a fundamentally different reality. It calls the man to get out of slavery of material circumstances, to the freedom of spirit.

But to finish the consideration of the effects of the intervention of the transcendent into the area of immanent on that pathos (but not losing verity) statement would be an obvious oversimplification of the problem. The process under consideration conceals a number of challenges.

First, I must say that it generates an internal conflict in a man. The essence of the conflict is reflected in the famous proverb about what you'd prefer: a real bird in hand or two in the bush. Dreams caused by the transcendental imagination are fascinating. But you cannot live only in the dream. As a person lives in material reality, he needs to ensure the immanent conditions of his live. Sooner or later, he is confronted with the harsh reality that becomes a source of conflict, acquiring, depending on the situation, either heroic or tragic character.

The choice is somewhere between the two extreme alternatives. On the one hand, there is a full freedom from material reality, contempt for it, and passing into Nirvana. On the other hand, submission of your life to the purely physical needs, oblivion of dreams, animal existence. The choice is either to go for the dream, destroying the familiar world, or to forget about it, which is extremely difficult. It is not surprising that some people hesitate a lifetime, unable to prefer anything definite.

Representatives of the Russian religious philosophy felt very well and expressed brightly the essence of this conflict. So, I. Ilyin wrote that some people remain spontaneous materialists because they take only that what is given them by the material nature and do not care about their spiritual side. They believe that it is enough to listen to your body and experience sensual sensations. With these sensations, they identify their "selves". They consider real only those things, which they perceive via sight, hearing, smell, and touch. For the rest, they pay no attention. Seriously, they perceive only experience, appearance, and external relations. They don't care what the inner life is (Ilyin 1994, p. 134).

Such people experience no internal conflict: the external data inside them just interferes with nothing. On the one hand, this condition looks very tempting. After all, these people are not tormented by a sense of responsibility, they are free of remorse, good and evil for them are only prejudices of the fools. Religious faith for such people is an empty tale. But, such a view leads to a flat world perception! Art for such people is only the means of exiting their sensory experiences. They seek only diversity, brightness, whatever tasteless it may be.

To illustrate this approach, I. Ilyin even wrote a small fairy tale. "In a distant country once there lived a cocky guy who wanted to know nothing about God. Then God punished him so, that everything he saw seemed only dead matter to him. No flower ever blossomed for him; no sacred image shined for him; and joy stopped singing in him. Motionless and poorly, empty and dead the dead matter smiled at him. Others could not understand him, being happy at flowers and blooming themselves with joy. Then he fell into despair and took his own life" (Ilyin 1994, p. 135).

Second, this process is a great temptation for a person. No one can guarantee that the pursuit of the ideal, beckoning by the pictures of the bright future, wouldn't turn into a complete collapse for enthusiastic people. Whether their dream is constructive in its basis or it is destructive, no one knows in advance. Following the transcendental, ideal is a challenge with all the unknowns. Will such an ideal enthrall a person deeper into the thicket, or will it lead to a wide open space, at the beginning of the path one cannot define.

What then? Perhaps, indeed, the best variant is to abandon such dreams, because it leads to some kind of suspicious, "otherworldly" reality, where familiar immanent laws and regulations stop to work? Without a dream, life is somehow simpler and easier. In the modern world this problem is often solved that way: if it is not possible to draw up rational reasonable representation of something, people often pretend that this something just does not exist. The results of applying this approach are obvious: the ubiquitous oblivion of higher values and ideals on the grounds that they do not withstand empirical validation. Material benefit and profit go ahead in people's motivation. Expelling the transcendental imagination for its violation of the borders of the immanent, we risk getting a pretty primitive world in which life is, at least, uninteresting.

But, regardless of whether or not a person wants it, he is always on the border, which runs between the transcendental and immanent imagination, he is always in a state of a very difficult choice. Otherwise, it is not about a human being, but rather about a supreme being, bored by human hesitations, or about the animal, which is indifferent to the sublime impulses of the soul.

References

Berdyaev N (1989) The meaning of the creative act. The philosophy of freedom. Pravda, Moscow, pp 252–580

Ilyin I (1994) I peer into the life. Collected works in 10 v. V.3. Russkaya kniga, Moscow, pp 89–226

Kant I (1994) Critique of pure reason. Myisl, Moscow

Solovyov V (1990) Justification of the good. Works in 2 v. V.1. Myisl, Moscow, pp 47–548

Part V
New Unity in Science and Humanities: Renewal of Ethics

Translation in a Hermeneutical Context: Transforming Culture and Human Nature Through Ethics

Andrei Kopylov

Abstract The paper looks at a number of problems connected with the ethical dimension of hermeneutical practices. It opens with a discussion of the opposition between the traditional approach to translation ethics which posits a 'code of ethics' for translators, and the approach put forward in a number of studies (V. Bibikhin, A. Berman, J.-L. Cordonnier, P. Ricœur, etc.) which envisage the translation phenomenon in its entirety. The author suggests that, whereas the former approach is narrowly deontological, the latter marks a turn towards virtue ethics, opening a new perspective on the problem of translation ethics. This turn implies a view of translation as a (trans)formative process which affects the recipient community's attitude to their own language and culture. The most fundamental virtue inherent in translation is identified as respect for the Other in their irreducible alterity. It is also argued that the idea of virtue ethics is applicable to other hermeneutical situations. This highlights the importance of the formative aspect of hermeneutical experience (H.-G. Gadamer). With ethics of difference and diversity as one of its guiding principles, hermeneutical experience may be expected to promote tolerance and respect for the Other.

Keywords Translation · Hermeneutics · Deontology · Virtue ethics

Introduction

Interest in the ethical dimension of translation is a relatively recent development in translation studies. As Theo Hermans points out, translation scholars traditionally approached their subject from a linguistic or a literary perspective and were mostly concerned with the textual dimension of translation, particularly with the relations between source and target texts; another traditional area of interest embraced such

A. Kopylov (✉)
Murmansk Arctic State University, Murmansk, Russia
e-mail: andreikopylov2010@yandex.ru

J. S. Methi et al. (eds.), *Borderology: Cross-disciplinary Insights from the Border Zone*, Springer Geography,
https://doi.org/10.1007/978-3-319-99392-8_17

practical matters as the training of translators and interpreters and translation criticism (Hermans 2009, pp. 93–94).

Translation, however, involves not only linguistic but also ethical and even political choices. It is obvious that the translator's position with respect to his clients entails certain moral obligations, and both the translators themselves, as a professional community, and society at large have long been aware of such obligations. In fact, the training of translators normally proceeds from a set of ethical principles which underlie many other aspects of the educational process in this area. It may be less evident, but can be shown, that other people involved in the process of translation, like the translator's clients, editors, publishers, etc., also find themselves in a position which is not ethically neutral. Finally, society at large has a certain view of translation practices which reflects its attitudes to contacts with other peoples, its sympathies and animosities, its treatment of the Other and its perception of alterity as such. All of these have a clear ethical significance. Being involved in the process of mediating between sometimes hostile communities, the translator finds himself vulnerable to their mutual hatreds and prejudices. An extreme and, by modern standards, somewhat exotic example of this vulnerability can be found in Plutarch's account of Themisticles' life. The legation sent by the Persian king to Greece 'to demand earth and water' was accompanied, among others, by an interpreter who duly did his job of translating the Persians' demands into Greek. Themistocles was so outraged by what he saw as 'prostituting' the Greek language in the process of expressing in it the demands of the Barbarians that he had the interpreter arrested and put to death. Plutarch recounts the episode with no obvious sympathy for the victim and notes that the whole affair was met with approval by the Greeks (Plutarch 1968, p. 19).

Here, the interpreter is held responsible not for what he has done in his professional capacity with the source message, but for the actual content of that message, which originated with someone else and over which he has never had any control. Classical Greek culture was prepared to treat him as a kind of accomplice to the 'Barbarian' king seeking to destroy Greek independence, which seems to suggest that his position was not perceived as that of a neutral intermediary (something we tend to take for granted when thinking of translators and interpreters nowadays). The Greeks were outraged by the demand for 'earth and water' to such an extent that everyone involved in the effort of getting it across to them was immediately classified as an enemy.

Classical Greek society must have held a view of the very phenomenon of translation which is so far removed from our own that we perceive Themistocles' behaviour on that occasion as completely absurd. Moral indignation, however, is normally the result of a feeling that some important ethical norm has been violated. What kind of ethical norm did the interpreter violate on that occasion, and did that norm have anything to do with translation as such? The answer to the latter question is probably 'no'. Classical Greek did not even have a special word for translation, and the practice itself was generally held in contempt, as it was normally performed by Hellenized 'Barbarians'—at that time, the Greeks themselves did not stoop to learning 'Barbarian' languages. This view of bilingualism and translation clearly entails a certain ethical position with regard to the treatment of translators and interpreters.

An analysis of the various types of translation ethics may help to reveal the significance of the ethical dimension of other interpretative practices. Translation is, quite obviously, a hermeneutical phenomenon. In fact, it is not uncommon for it to be treated as a paradigm of hermeneutical practice in general since in the case of translation certain aspects of hermeneutical experience become easier to perceive than in other interpretative practices. Hans-Georg Gadamer remarks that the conditions of understanding, in general, are best revealed in those situations where it is not achieved easily, where it seems to have encountered some kind of obstacle. A good example of this type of situation is afforded by a conversation between people who speak two different languages and, therefore, need to have their conversation translated by someone who knows both these languages. The meaning of an utterance in one language must be preserved in the other, but it also has to be transplanted into a new context created by the target language and culture, which may sometimes make it difficult for the translator to keep the new utterance both readily understandable to its recipient and sufficiently close to the original message in terms of reproducing its exact meaning. Therefore, every translation necessarily involves interpretation; in fact, Gadamer goes as far as saying that the translation of somebody's words into another language represents the climactic point of the interpretation that the translator has made of those words (Gadamer 2004, p. 386).

To some extent, this is due to the trivial fact that in translation the result of the interpretation it involves must invariably take the shape of a separate text which makes the translator's understanding of the original explicit. Like anyone who happens to have read the original text or heard the words of a speaker in a conversation, the translator may find that his source text is less than perfectly clear on this or that point, but he is not generally entitled to leave his translation equally ambiguous, he has to commit himself to one of the possible interpretations, to show how exactly he understands the original message (Gadamer 2004, p. 388).

In this paper, I will attempt to show that the ethics of translation is something that pertains not only to translators themselves, but to entire communities which rely on translations for some of their cultural practices, that the ethical position adopted by a linguo-cultural community with respect to translation may seriously affect its cultural development, and that embracing an ethics of difference and diversity may be a way of not only culturally benefiting from translation, but also of perfecting human nature through hermeneutical experience.

It seems reasonable to begin this discussion with what is perhaps the most traditional view of the ethics of translation.

The Ethics of Translation as Deontology

The most common approach to the problem of the ethics of translation seems to be a purely pragmatic one. Like many other professions, that of the translator has its own 'code of ethics' which may vary from country to country and may or may not take the form of an officially recognized document, but which is always focused on the figure of the translator or interpreter to whom it prescribes a certain set of rules and principles designed to make them aware of the duties they have towards their clients in the context of their professional activities. Let us consider the example of the *Translator's Moral Code* contained in the standard Russian textbook *Introduction to Translation Studies* by Irina Alekseyeva. Her Code is reasonably compact and appears to comprise the most important principles that tend to be included in such documents. Though it is by no means legally binding, it is generally respected, and its principles are largely followed by the professional community in Russia. In other words, Irina Alekseyeva seems to have brought together what has come to be recognized as the fundamentals of the translator's professional ethics.

The principal points of this Code may be summed up as follows. (1) The translator is, above all, a mediator whose task is to express the message and contents of the source text in the target language. He is not, therefore, a participant in the act of communication in which he is involved in his professional capacity. (2) The source text must be treated as inviolable by the translator who may not introduce any changes into it, e.g. it may not be abridged or expanded in the process of translation, unless the translator has been specifically commissioned to do so. (3) It is the purpose of the translator to render the content and other significant features of the source text as fully as possible in the target language. (4) The translator may not infringe upon the rights of his clients. (5) On special occasions, such as international contacts between senior representatives of governments, the translator may be required to perform certain diplomatic functions which involve efforts to prevent unnecessary tensions between the countries concerned. In this case, it becomes ethically acceptable, and indeed desirable, to slightly modify such linguistic, stylistic or cultural features of the original message as are likely to cause unintended offence and thereby create or increase tensions between the two parties. If this is not the case, the translator may not interfere in the relations between the parties involved or in any way reveal his own attitude to the content of what he has been commissioned to translate (see the complete Code in Russian in Alekseyeva 2006, pp. 29–30).

Understandably, the above 'code' is focused entirely on the figure of the translator. Other participants in the process of mediated cross-cultural communication are barely mentioned, and when they are, it is only to emphasize that their rights are not to be infringed upon. In the absence of a similar 'code of ethics' for anyone else taking part in this process, one is tempted to ask if the translator is, in fact, the only person involved who is expected to take any ethical responsibility for the smooth running of the process at all. This point will be discussed later in this paper. Another thing that almost immediately strikes the eye is that almost all the provisions in this code can easily be traced back to the traditional view that the translator must be 'invisible', as far as possible. Indeed, this code may be described as a code of the 'translator's invisibility' (cf. the title of L. Venuti's famous book). Whatever in the code cannot be reduced to the idea of the translator's invisibility is treated as a special occasion, as something that is, in a sense, a deviation from the translator's normal role.

It is also obvious that this code is an example of *deontological* ethics, which would, of course, appear to be natural for a code of professional ethics. In some countries, in fact, the word 'deontology' itself tends to appear in the titles of such documents, cf. *Le Code de déontologie de la SFT* (SFT stands for the 'Société française des traducteurs'). But is deontology the only kind of ethics that is relevant to translation? Can it possibly be the case that this role of the translator's professional deontology is merely a consequence of the fact that, when speaking of the ethical dimension of translation, we simply tend to overlook the role of everyone else? After all, it would hardly be fair to say that the translator is the only one whose position with regard to a translating act entails some form of moral responsibility.

The Ethics of Translation as Virtue Ethics

Those who commission a translation, or take part in a discussion speaking through an interpreter, or merely read a translated text have to relate in some way to everyone else involved in the translating act in question and even to the source and target texts. This kind of relation does seem to have an inherent ethical dimension inasmuch as it entails a spectrum of choices on their part which may affect the outcome of the act of communication and, to some extent, even the well-being of some or all of the people concerned.

For instance, when speaking through an interpreter, one may ignore the fact and choose to speak in exactly the same way as one would to someone of one's own cultural and linguistic background, leaving it entirely to the translator to make the target message acceptable to its recipient; on the other hand, one may take into account, as far as one can, the difference of cultural and linguistic contexts of the two messages and choose to speak in a way that is reasonably free from unnecessary idiomatic expressions, puns or references to those elements of the source culture that might be incomprehensible to the other party and/or difficult for the interpreter to render adequately in the target language. To some extent, this is simply a matter of expediency and common sense: if you wish to succeed in communicating with someone who does not speak any of the languages you know, you must realize from the very outset that, as a speaker, you are only as good in his perception as your interpreter is, and that you would do well not to burden your interpreter with unnecessary linguistic and cultural difficulties that he may find himself unable to overcome. But at the same time, this is also an ethical choice affecting your interpreter and your foreign partner—are you prepared to go some way towards meeting their specific needs, if it means making your own situation somewhat less comfortable, if it involves some extra effort on your part in formulating what you wish to say?

Even when we read literary texts in translation, we have to relate to them in a way which is not ethically neutral. What do we expect the translator to have done? Do we expect him to have rendered the source text in a fluent, easy-to-read manner, eliminating or adapting those elements of the original that require some form of commentary and making the style of the translation resemble that of an original work of our own literature? And if we find that he has not done so, will we be prepared to make an extra effort that may be required to work through those parts of the text and commentary which are strange or difficult to understand, by the standards of our own culture? Are we prepared to accept the 'otherness' of a translated text, to concede that the familiar forms of our culture are not necessarily the only possible ones, that it may be a good idea to borrow something from the source culture into our own by means of translation, rather than seeing the latter as a mere transfer of universal meanings? The moment we stop to think about any of these or similar questions, we begin to develop an ethical position on translation, even if we have not translated a single sentence in our entire life.

In the past 30 years or so, a number of papers and books have been published which deals not so much with the translator's professional ethics as such, but rather with the ethical foundations of translation as a specific form of interlingual and intercultural traffic. Their authors have something in common: their interests lie outside the linguistically oriented paradigm of translation theory which has dominated translation studies since the mid-twentieth century in several countries, including Russia, though it has never enjoyed similar prominence in the English-speaking world. In this paper, I mostly draw on the work of Vladimir Bibikhin, Paul Ricœur, Henri Meschonnic, Antoine Berman, Jean-Louis Cordonnier and Lawrence Venuti, although the following discussion focuses on the views of only some of them.

Vladimir Bibikhin, a highly original Russian philosopher and translator of a number of key works of Western philosophy, including an acclaimed Russian version of Heidegger's *Sein und Zeit*, did not normally use the term 'ethics' to refer to the aspect of translation practices which is under discussion here, but in his short article *Podstanovochny Perevod [Substitutive Translation]* (1973) he took up the concept of 'hospitality' with reference to translation, which does appear to be an ethical position. According to Bibikhin, no effort to introduce into the translating culture a new text originally written in another language can be expected to succeed, unless the translator comes to realize that he must, above all, act in the spirit of hospitality, treating the work to be translated as a welcome newcomer. A long process of trial and error will teach the translator that a welcoming, hospitable attitude is a necessary precondition for enabling the author's thought to find its way into a new language and culture (Bibikhin 2001, pp. 201–202). It is noteworthy that, on Bibikhin's view, this ethical position has to be *discovered* by the translator in the process of translation, unlike the rules of professional deontology which, from the perspective of an individual translator or interpreter, are imposed by some extraneous authority before the translating act itself takes place.

The idea of linguistic hospitality is also prominent in Paul Ricœur's writings on translation which offer a highly unconventional perspective on the subject. Ricœur treats the problem of translatability as that of the commensurability of languages. He criticizes the commonly accepted view, according to which translation deals with a kind of preexisting meaning which has to be extracted from the source text and then verbalized in the target language. If that were the case, the meaning to be rendered in translation would be conceivable as a 'third text' mediating between the source and target texts, against which the accuracy of a translation might be judged. According to Ricœur, this is not the case: meaning emerges in the process of translation and may even be described, in a sense, as its product. One of the reasons why we normally fail to see this is that all the European languages have long been in constant contact with each other, and many of them are also related to each other. The kind of translatability that exists between the languages of Europe and that we take for granted is entirely a matter of genetic relationships between them and of their long-standing cultural contacts. It is when we encounter a language and culture radically different from our own that it becomes clear that translation can rely on no clear-cut preexisting meaning easily comprehensible to the target culture, and that it ultimately has to deal with the problem of incommensurability. In such situations, it is the task of translation to create, or build up, the commensurability that was initially not there. It does so by giving old words subtle new meanings, by introducing unusual combinations and borrowing words from the source language. Gradually, the two languages and cultures become 'comparable', something that has repeatedly happened in European history, e.g. Luther's translation of the Bible made German a language commensurable with Hebrew and Greek (Ricœur 2004, pp. 53–67). To achieve its aims, translation requires a certain ethical position: the translator must be able to exercise 'linguistic hospitality' which consists in being able to welcome the foreign into your own language and culture. This does not have to be seen as a form of sacrifice on the part of the target culture or the translator; the latter can, in fact, find happiness in extending hospitality to another language, even when his work acquires a dramatic dimension due to the agonistic element inherent in his task (Ricœur 2004, p. 19). In a sense, on Ricœur's view, the ethical position of linguistic hospitality is a condition of possibility of the kind of translation which can create commensurability.

Some of Ricœur's key views echo, to an extent, those of Antoine Berman of whom he wrote with nothing less than admiration. Not unlike Ricœur, Berman takes it to be the principal ethical aim of translation to receive 'the Foreign as Foreign' (Berman 2000, pp. 285–286). However, historically, literary translation has been dominated by an ethnocentric approach which seeks to adapt the Foreign to the norms of the recipient culture as far as possible, leading to a number of 'deforming tendencies' that continue to plague translation almost everywhere in the West and, maybe, elsewhere in the world. These tendencies are interconnected and can, in fact, be said to function as a system, although some of them are better known to the general public than others. Some are more pronounced in a particular literary tradition, but fundamentally they are common to the entire Western world (Berman 2000, p. 288).

Berman discusses 12 deforming tendencies, although he says that there may be more of them. To give a few examples of how they work, one might turn to some of these tendencies and consider their effects. *Rationalisation* mostly affects the syntactical organization of the text; it means that the translator tries to reconstruct parts of the original with a view to imposing on the text a kind of rational order which he takes to be a necessary discursive convention, no matter whether it is observed in the source text or not. *Ennoblement* aims at achieving a level of stylistic elegance which is required by the recipient culture's standards of poetic expression or good prose writing. This attitude reduces the original text to the status of raw material which the translator turns into an elegant finished product. *The effacement of the superimposition of languages* consists in making linguistically homogeneous those texts that in the original are characterized by a complex interplay of different linguistic elements such as the literary standard and dialects, different forms of a language (e.g. Castilian and Latin American Spanish), etc. The effect of this tendency is to diffuse the linguistic tensions that underlie the unique style of the original, making the translation bland and impoverished (Berman 2000, pp. 288–296). Even these three examples are sufficient to show that the reader of a 'deformed' translation will not be in a position to appreciate the foreignness of the foreign text, to experience it in its alterity. The deforming tendencies, however, have endured for centuries without being challenged. That would seem to suggest that there must be some deeply rooted cultural mechanism behind them.

On A. Berman's view, the reason for the existence of these 'universals of deformation' (Berman 2000, p. 296) is the fact that translators have assimilated what he calls 'the Platonic figure of translation'. The latter presupposes a number of oppositions such as sense and word or spirit and letter, which, when applied to translation, result in giving indisputable priority to meaning over the letter. The tendency to express the supposed meaning of the original in the clearest and most elegant way possible devalues the peculiar style of the original, its letter and makes ethnocentric translation appear thoroughly natural (Berman 2000, pp. 296–297). Its deep cultural roots mean that it has not often been considered from an ethical perspective. But Berman sets out to do just that and suggests an ethically motivated alternative to the 'Platonic figure', viz. a kind of translation which respects the letter and the 'signifying process' of the original, rather than its bare meaning. He maintains that working on the letter of the original is in some important ways a more fundamental task of the translator than re-wording its meaning in another language. By giving the letter its due the translator not only reproduces the signifying process of the original, something more profound than its meaning but also contributes to the transformation of the target language. This shows that translation plays an important 'formative role' in culture (Berman 2000, p. 297).

Thus, taking Berman's view a stage further, one might say that the deforming tendencies in translation have been able to exert their destructive influence on translators, as well as on cultural exchanges in general, for the simple reason that the ethical dimension of translation has customarily been ignored in most discussions of translation. Before setting out to translate something culturally important, one ought to be completely aware of the ethical aim of translation which is to experience the Foreign in its otherness, in its irreducible difference from the norms and practices of the translating culture. This ethical aim transcends the limits of professional deontology, as it presupposes that the reader, too, ought to be prepared to make a kind of hermeneutical effort which is required if one is to understand something that does not conform to the accustomed cultural forms and values. In fact, if Berman's 'figure of translation' is to be realized at all, the ethical aim that he has formulated must be recognized by the entire translating culture which would have to accept a 'foreign influence' of the kind that many cultures desperately try to avoid. Berman's 'figure of translation' requires more than just a new type of translator, it requires a new public and a new cultural and moral climate.

Criticism of ethnocentric translation may also be found in the work of Cordonnier (1995), who suggests a return to Schleiermacher's ideal of taking the reader to the author as an alternative to the current tendency towards cultural adaptation, and of Venuti (1995) who analyses the literary, cultural and social implications of 'fluent translations' and 'the translator's invisibility' in the English-speaking world.

The approaches to the ethics of translation found in the work of Bibikhin, Ricœur, Berman and Cordonnier have something in common—they are all versions of *virtue ethics*, rather than professional deontology, so they represent a significantly different type of ethics. With reference to translation, virtue ethics seems to presuppose the acquisition of certain personal qualities, such as 'linguistic hospitality' or the ability to accept the Foreign as Foreign, which may be seen as *virtues* enabling one to achieve a certain desirable quality of translation, such as fidelity to the original, or to enjoy a translation which is not 'fluent'. It is a very important feature of this type of translation ethics that it is not confined to the figure of the translator, but may, in fact, be extended to the entire recipient community. It is equally important that, if one were to interpret the relevant writings of the authors mentioned above in terms of virtue ethics, the most fundamental 'virtue' required of those concerned with the translation phenomenon would have to be identified as respect for the Other in their irreducible alterity, and this particular kind of ethics would also have to be defined as *ethics of difference and diversity*.

Virtue and Interpretation

Translation ethics as virtue ethics opens up a new perspective on translation as such. Here translation appears to be *more than mere mediation,* which is the case with the deontological perspective on the ethics of translation.

In the framework of virtue ethics, translation has to be conceived of as a *(trans)formative process* which effects profound changes in the subjects involved in it, both individual and collective. At the individual level, this is a matter of forming habits. Aristotle described 'states of character' as products of engaging in 'like activities', emphasizing the importance of forming the right kind of habits from one's youth (Aristotle 2009, p. 24). At the collective level, as Paul Ricœur has shown, translation serves to transform translating languages and cultures, renders them comparable, commensurable with others, introducing them in this way to global culture.

However, there appears to exist a wider hermeneutical framework for virtue ethics. At the beginning of this paper, it was mentioned that in hermeneutics translation is sometimes treated as a paradigm of interpretation as such. This view seems to justify the idea that virtue ethics may be extended to the entire sphere of interpretative practices.

The (trans)formative significance of translation revealed in the perspective of virtue ethics finds what appears to be a close parallel in Gadamer's concept of *Bildung*. A useful interpretation of this concept is suggested by Chris Lawn and Niall Keane in their *The Gadamer Dictionary.* As the relevant entry in this Dictionary shows, the word *Bildung* is one of those German philosophical terms which are difficult to translate into other languages. It is important for our purposes that in its normal German usage this word may mean 'formation', 'culture' and 'education'. With Gadamer, this term serves to denote a specific way of conceiving the fundamental hermeneutical process of understanding which has its roots in the premodern idea of 'humanistic scholarship'. In this sense, *Bildung* is opposed to *method.* The latter is recognized by modern science as 'the correct procedure for the accumulation of facts and knowledge', normally in the domain of natural sciences. *Bildung,* on the other hand, is associated with the humanities. Unlike natural sciences, the humanities cannot achieve their purpose by methodically gathering facts, formulating hypotheses and testing them in a series of experiments. They are concerned with culture which is the product of 'organic development' and which functions differently at the individual and the social level. For an individual, *Bildung* is largely a matter of socialization, it is a process in which the person gradually acquires social habits, customs, traditions and other elements of a particular society's cultural heritage. At the level of society, *Bildung* describes the process of the historical development of a culture which involves gradual transformations of its customs and traditions (Lawn and Keane 2011, pp. 17–18). Gadamer remarks that in the nineteenth century, the earlier meaning of the word connected with various natural forms became almost separate from the new meaning connected with culture that it had recently acquired, notably in Herder's writings. From that time on, *Bildung* began to be treated as an important term in the German vocabulary describing various aspects of culture. Now it is used to denote the way people develop their 'natural talents and capacities' (Gadamer 2004, p. 9).

The idea that *Bildung* is a way of developing people's 'natural talents and capacities' introduces us to an important feature of interpretative practices, viz. to their being ultimately rooted in human nature. In fact, the 'homogeneity of human nature' was sometimes taken to be the reason why understanding is possible in the first place, a view held, among others, by Wilhelm Dilthey (Gadamer 2004, p. 226). The notion of human nature is equally important in the context of virtue ethics in the sense that the acquisition of virtue is made possible by some basic predispositions of human nature, but, at the same time, requires habituation. In other words, we are naturally capable of acquiring virtues through the habitual exercise of virtuous feelings and behaviour (Aristotle 2009, p. 23). It is an important consideration for our purposes that both Aristotelian ethics and Gadamerian hermeneutics seem to treat cultural phenomena as being rooted in certain basic aspects of human nature. This is a position that not only overcomes the nature-culture divide but also leaves room for perfecting human nature by means of cultural practices.

The twofold character of *Bildung* as both a way of acquiring a certain culture and the process of that culture's historical development suggests that there exists a form of mediation between the two levels of *Bildung* which renders the individual ethical choices made in the context of *Bildung* a matter of more than merely personal significance. We contribute to the culture that we acquire by the very fact of acquiring it, and the way it develops from now on is to some extent dependent on the sum total of our individual interpretations of what we have acquired. We are ultimately responsible for the way we relate to the cultural contents and values that we assimilate in the process of *Bildung* and for the 'virtues' that we acquire in its course. In this respect, the experience of reflection on the social and cultural practices of translation appears to contribute important insights into the ethical dimension of the matter. The idea that the very feasibility of an adequate translation is dependent on the ethical positions of both translators and recipients and that the kind of position required is one of acceptance of difference and diversity appears to be applicable to hermeneutical experience at large. The ethics of difference and diversity makes for the acquisition of the virtues of openness and tolerance which are required in any case of interpretation if it is expected to reveal the Other in his or her alterity. The latter, of course, is by no means the only possible purpose of interpretation. In fact, the question of the purposes of interpretation is one where hermeneutics has to meet a host of external considerations, some of them ethical.

For an example of such external considerations, let us turn to translation again. The discussion of the ethical dimension of translation in the work of Bibikhin, Ricœur, Berman and Cordonnier aims at establishing a kind of ethical framework for translation 'from within' the translation phenomenon itself; it seeks to discover an ethical basis for translation which is inherent in the practice itself. Translation, like most other interpretative practices, however, can be influenced by social, cultural and political considerations that, in effect, prescribe an ethical framework for it 'from without'. The so-called Brazilian 'cannibalism' is a good case in point. The proponents of this school of translation use the metaphor of 'cannibalism' to denote a new type of translation (and, therefore, interpretation) which treats the original belonging to one of the former colonial cultures as a source of 'nourishment' for the translating culture. By appropriating a text written in the highly cultivated language of the former colonizer the post-colonial recipient community acquires new energy and vitality 'at the cost' of their former colonizer (Munday 2001, pp. 136–137). Here, translation becomes a method of obtaining a cultural 'compensation' for centuries of colonial control. It is obvious that this kind of hermeneutical practice is based on a set of ethical principles which have little or nothing in common with the ethics of difference and diversity discussed above.

Finally, the translator can find himself in a situation where he feels that all normal ethical considerations pertaining to this hermeneutical practice have to be abandoned in the name of far higher moral values. Erich Prunč discusses the following situation of this kind. Under the Nazi regime, Margarete Buber-Neumann, who had to act as interpreter in the KZ Ravensbrück, used her position to help other prisoners by distorting in translation what they were saying to their interrogators, so that they would not be killed by the Nazis (Prunč 2005, p. 173). By doing so she put her own life at risk. She violated every conceivable principle of translation ethics, but her behaviour is seen by us as worthy of the highest praise possible. In this case, the true object of moral judgement is not her performance as a translator, but her 'act' as a human being, something that Mikhail Bakhtin might have described as a *postupok* (Bakhtin 1993).

Conclusion

Our discussion of the opposition between the traditional approach to translation ethics which focuses on a 'code of ethics' for translators, and the approach put forward in a number of studies of the ethical dimension of translation viewed in its entirety (V. Bibikhin, A. Berman, J.-L. Cordonnier, P. Ricœur, etc.) was undertaken with a view to showing that, whereas the former approach is *deontological,* the latter marks a significant turn towards *virtue ethics*, opening an entirely new perspective on the problem of translation ethics. Since virtue ethics is ultimately concerned with fundamental features of human nature, the significance of this turn in the understanding of translation ethics lies largely in the fact that it implies a view of translation as a (trans)formative process which affects the recipient community's attitude to their own language and culture, and enriches the expressive potential of the target language. In this context, the most fundamental 'virtue' inherent in the translation phenomenon ought to be identified as respect for the Other in their irreducible alterity, defining this type of ethics as *ethics of difference and diversity*. In a wider hermeneutical context, the idea of virtue ethics appears to be applicable to other hermeneutical situations, in fact, to almost any kind of interpretative activity. This would seem to highlight the importance of the formative aspect of hermeneutical experience (cf. H.-G. Gadamer's notion of *Bildung*). With ethics of difference and diversity as one of its guiding principles, hermeneutical experience emerges as something that may be expected to promote tolerance and respect for the other in their alterity.

References

Alekseyeva IS (2006) Vvedeniye v perevodovedeniye (Introduction to translation studies). Academia, St. Petersburg, Moscow
Aristotle (2009) The Nicomachean ethics (trans: Ross D). Revised with an Introduction and Notes by Lesley Brown. University Press, Oxford
Bakhtin M (1993) Toward a philosophy of the act (trans: Liapunov V). University of Texas Press, Austin; Russian edition: Bakhtin M (1986) K filosofii postupka. In: Filosofiia i sotsiologiia nauki i tekhniki: Yezhegodnik 1984–85. Nauka, Moscow, pp 82–138
Berman A (2000) Translation and the trials of the Foreign (trans: Venuti L). In: Venuti L (ed) The translation studies reader. Routledge, London and New York, pp 284–297; French edition: Berman A (1985) La traduction comme épreuve de l'étranger. In: Texte, n° 4, pp 67–81
Bibikhin VV (2001) Slovo i sobytiye (The word and the event). Editorial URSS, Moscow
Cordonnier J-L (1995) Traduction et culture. Les Editions Didier, Paris
Gadamer H-G (2004) Truth and method. 2nd, rev edn (trans: Weinsheimer J, Marshall DG). Continuum, London, New York
Hermans T (2009) Translation, ethics, politics. In: Munday J (ed) The Routledge companion to translation studies, Rev edn. Routledge, London and New York, pp 93–105
Lawn C, Keane N (2011) The Gadamer dictionary. Continuum, London, New York
Munday J (2001) Introducing translation studies. Routledge, London
Plutarch (1968) Plutarch's lives with an English translation by Bernadotte Perrin, vol II. Harvard University Press, Cambridge, MA; William Heinemann Ltd, London
Prunč E (2005) Translationsethik. In: Sandrini P (ed) Fluctuat nec mergitur. Translation und Gesellschaft. Peter Lang, Frankfurt/Main, pp 165–194
Ricœur P (2004) Sur la traduction. Bayard, Paris
Venuti L (1995) The translator's invisibility. A history of translation. Routledge, London and New York

The Need for Being Disinterested as a Key Characteristic of Human Nature

Andrei Sergeev

Abstract The article focuses on investigating the human need for performing fundamentally non-utilitarian actions and on people's orientation towards being disinterested whereby they are introduced to, and partake of, the whole. The actualization of our life is connected with the human ability to act and with our performing deeds, when we associate ourselves with a principle and thereby undertake the labour of collecting ourselves. In this context, it is essential to turn to the ultimate and the border, since they are connected with our need for expending ourselves and abandoning ourselves to the other, when an extended perception of life as a gift (an act of giving) is defined and the other turns out to be included into our understanding. The author discusses the meaning of the act of symbolization and analyses the symbol and the sign in the context of the human being's engagement with sense and reasoning. Special emphasis is given to the idea that the human need for performing disinterested acts becomes more acute with the weakening and disappearance of a presumption according to which any moment of life can, in principle, be consciously comprehended, whereas consciousness itself is perceived as an unlimited power enabling the human being to associate himself with anything whatsoever.

The Need for Being Disinterested

Alongside utilitarian actions directed at maintaining our existence, we sometimes also perform non-utilitarian actions. Moreover, a need periodically manifests itself for a person to perform disinterested actions, which can be assumed to be somehow connected with our—purely human—nature.

One can, of course, try and explain disinterested actions by resorting to a deterministic description. In fact, such 'explanations' are suggested from time to time: we find it difficult to accept disinterestedness, proceeding as we do from the

A. Sergeev (✉)
Murmansk Arctic State University, Murmansk, Russia
e-mail: asergeev8@yandex.ru

J. S. Methi et al. (eds.), *Borderology: Cross-disciplinary Insights from the Border Zone*, Springer Geography,
https://doi.org/10.1007/978-3-319-99392-8_18

cares of our existence which are focused on solving utilitarian tasks and are penetrated by cause-and-effect explanations. As far as accepting the 'purity' of our actions and thoughts is concerned, everything depends on ourselves, or rather—on our orientation or disorientation in *relation* to that which is disinterested. However, from time to time, we do experience, if not a need to commit such actions ourselves, then, at least, an urgent need to be somehow involved in them.

As an answer to the questions why and for what purpose this occurs, one can suggest the following: it is something in which the human need for purification is manifested, a need for severing one's connection with the particular and for replenishing what one has lost as a result of one's internal 'expenditure'. If we do, after all, accept the disinterested as such and do not make desperate attempts to discover a certain mercenary motive behind what is being done, seeking to relegate it in this way to the category of utilitarian actions, and then it is connected with an intention to distance ourselves from all things *particular* and *contingent*. We find it a matter of vital necessity that we—either directly or indirectly—break free from particular and, generally speaking, contingent actions, which means concentrating in some way on the *whole*.

The Particular and the General. The Particular and the Whole

It is important to emphasize here that when utilitarian actions predominate, the exit from the particular is connected with the *general*. On the contrary, a person's need for the disinterested discloses another aspect of their life connected with partaking of the whole. It should be noted in this regard that the general cannot deliver us from the particular: an orientation towards the particular can only be overcome by the whole, or rather, by our connectedness with it.

Let it be noted that we usually can see things only either as a whole or as a particular, and if we can see the one, we cannot see the other: the one eclipses the other, though we can remember what we have encountered before. At the same time, we should not lose sight of the fact that the very *ability to notice* both the whole and the particular is inherent in every single person. What matters is the fact that we *can* do it. Alongside being able to say that someone is overweight or slim, generous or avaricious, we are capable of perceiving them as a human being in their integrity. The same is true of life as such: we are essentially able to fix our gaze on something separate, but we can also see a whole life in one of its moments, and we can see bread in breadcrumbs.

But then, it is our attitude to the particular that makes us see dirt, rather than earth on the soles of our shoes. In the vast majority of situations this attitude, focused as it is on the particular, comes out on top and relegates a holistic attitude to life to the periphery of life. A *mediated* attitude to life that manifests itself in the functional, technical, technological and instrumental types of this attitude cancels out our *immediate* attitude to it. It is easy to see that in modern life, a practical attitude to things has become universal, whereas a principled non-utilitarian attitude is perceived as something that 'has been overcome' in a historical sense.

The experience of a holistic understanding of life in each of its moments is difficult precisely because of the *temporality* of our life. It is difficult to reveal the immortality of life, proceeding from the death of a human being and within the limits of experience connected with a sense of one's 'temporariness', because it often results in vanity, willfulness and arbitrariness, and in the conditions of the loss of a holistic perception of life this induces us to turn to *social instrumentalism*. Since our perception of the integrity of something is changeable, it is important to pay attention to any experience of our 'collision' with the whole. The ontological experience of comprehension of life is another dimension of the experience of human life, as compared with the experience of its contensive dimension. It is the *other* of our life.

Aristotle's 'Metaphysics': The Problem of Movement and Actualization

Human existence is impossible without performing actions: our existence manifests itself in actions and, one might say, is 'nourished' by them. However, as Aristotle has it, action can be movement but it can also be 'actualization' ('activity'). And if existence is realized through movement and by performing movements, then the entity manifests itself precisely in actualization. Being manifests itself in activity, when, along with movements, thinking reveals itself and connecting as it does different movements.[1]

Actualization as Collecting Ourselves in an Orientation Towards a Principle

Thus, human nature is connected precisely with *actualization*, and actualization can be connected with performing *disinterested actions*—with performing *deeds* (or acts—*postupki* in Russian). It would seem that it is the shortage of deeds that we consistently attempt to compensate for with our *will*, as we seek to add it to almost every action of ours. Incidentally, here in Russia nothing, no business at all is now solved without recourse to will. However, it remains questionable whether we can

---

[1]Aristotle. Metaphysics. Book IX. Chapter 6. In A. V. Kubitsky's 1935 Russian translation "entity" [*sushcheye*] is connected with "activity" [*deyatelnost'*]. See: Aristotle. *Metafizika* [Metaphysics]. Moscow and Leningrad, 1935, pp. 155–156. In the edition of this translation published in 1975 by M. I. Itkin "entity" manifests itself in "actualization" [*osushestvleniye*]. See: Aristotle. *Metafizika* [Metaphysics] in Aristotle, *Sochineniya v 4-kh tomakh* [Collected Works in 4 volumes], vol. 1. Moscow, 1975, p. 242. The Greek word is ἐνέργεια (1048b-35).

replace other things: for instance, whether will can compensate for a lack of disinterested action.

In performing a deed, people come to discover themselves from an unfamiliar and unaccustomed angle—they discover themselves from the position of a *principle*. It is a matter of considerable significance to a human being to be able to develop a principled attitude to any of their feelings, any of their acts of thought or will, even if such an attitude is not actually given, but only posited. Here we seem to be dealing with the human need 'to stand on the edge': 'on the edge' both of oneself and of what one is associated with. Being bound up with the 'disinterested principle' is something that actually allows us to *collect* ourselves, literally 'extracting' our being from particulars.

In other words, by performing disinterested actions and correlating themselves to a principle, people distract themselves from particulars and form a bond with themselves, coming face to face with themselves. In such actions—those of actualization—people treat themselves as beings who should be accepted the way they are. But it is here, in fact, that the possibility of rectifying ourselves has its source.

Singularity Points: The Possibility of Communication with the Ultimate
Actually, we are dealing with a *singularity point* where the 'advantages' and 'shortcomings' of our life so far are reduced to zero and a rift with the definiteness of our existence is made possible. At such a point, a change of the existent is possible: here the very orientation of the human being towards the *implicit* is important, as his or her action is tested regarding what might be described as its '*event* worthiness'. To put it very simply, at this point our action can or cannot be understood—by ourselves and by others—as an event.

At the same time, it is clear that a disinterested action appears to be a threat to everything that has already happened and taken place, i.e. a threat to the existent. Here, we can also see a hint at the fact that the essence of the human being is not confined exclusively to the preservation and reaffirmation of his or her existence: each of us needs to find a way to connect himself or herself with the *ultimate*. And if we fail to correlate ourselves periodically to the essential and to break through the fabric created by our utilitarian actions directed at the maintenance of our existence, then we run a serious risk of sliding into the boundless. And in view of this slide of ours into the boundless it cannot but be perceived now in an exclusively negative way.

Enthusiasm and Self-abandonment

When we are *keen* on, or *enthusiastic* about, something, we *abandon* ourselves to it. We become keen on, or enthusiastic about, some *business*, meaning business per se which can take any form whatsoever. This must be a kind of business that we abandon ourselves to beyond measure.

However, it is only someone who *is* that can abandon himself to something. On the contrary, when one is entirely beyond oneself, i.e. when one 'is' in a miscellany of things, one cannot abandon oneself to anything. If a person is beyond himself, then he is unable to abandon himself to another: it can only be done by someone who is 'collected' in his being. It is the ultimate that collects and concentrates our being, thereby capturing it for the things we are keen on.

Enthusiasm for some business, even one not connected with thought, but rather with passion or even a game, is something we are capable of apprehending and accepting in principle, even if are not inclined to justify it. We understand that what has captured a person is stronger than him. We understand that the person is being tested by a force which he cannot resist and, therefore, has to accept.

What we need is to learn to accept this force as such, i.e. to accept it as it is, whereas we are normally inclined to perceive it rather as something that exerts an external influence on us. We ordinarily perceive it as something that changes the contours of our existence, i.e. we consider the force which is carrying us away together with the impression which it produces on us and therefore we cannot actually distinguish our understanding of this force from the image or idea we have of it. We find ourselves dependent on this force and even come to perceive ourselves as its 'products', since our idea of this force anticipates our meeting with it in reality. And we come to feel a certain *fear*, we become afraid. We are now afraid of any possible meeting with a force capable of carrying us away in a wave of enthusiasm.

We find it extremely difficult to accept it *as* it is and take it for *what* it is. For this purpose, it is necessary to look at it directly, which means to see it with our own, and not with other people's eyes, it is necessary not to be blinded by either the force, or other people's opinions of it, and it is also necessary to be able to connect oneself with one's thought to see such a meeting as a chance that has come your way. A chance of meeting yourself.

It is noticeable that a meeting with what carries people away in a wave of enthusiasm is always linked to a sense of heightened inner energy, as well as to a change in their mood, as they face an inflow of energy which is superior to them and which, therefore, brings them a sense of being able to extend the boundaries of what is possible well beyond their normal scope. In their existence, a gap suddenly appears, and, finding their way into it, people acutely experience a need for actualization.

Energy: Storing and Expending It

In maintaining our existence, we are guided by a perceived need to store as much energy as possible and to discover a variety of sources for its replenishment that could provide a constant *growth* of its resources. On the contrary, actualization is connected with an extreme concentration of energy and its maximum *expenditure*, for we understand that it is a way of realizing our essence: in the matter of expenditure, we actualize ourselves in all our integrity. We experience an inflow of energy, as we are guided by the power of the *whole*. And as any definiteness is connected only with the circumstances of our movement, it is easily overcome. In this case, we are prepared to easily sacrifice any definiteness of ours. The event of meeting a *power* that captures us becomes an event of man meeting the *integrity* of himself.

Thus, our need for partaking of the boundless cannot be considered exclusively in terms of formulating the questions 'Why do we do it?' or 'Why does it occur?', but, on the contrary, it has to be connected by us with the need for our *completion* and *replenishment*. Acceptance of what expands and strengthens us, i.e. what fills us to completion, is more important in principle than we ourselves are, we who are aware of ourselves within purely individual limits.[2] Turning to what is disinterested and uncaused, we find in ourselves that which pertains to the *other*. We find that which is inherent in us, but which we do not know in ourselves yet.

Disinterested Acts and the Problem of Actualization

Disinterested acts enable us to live, growing up and outgrowing ourselves, which should at the same time be interpreted as *finding our way to ourselves* and to *our actualization*. It is important to note the forms which help us to retain the disinterested attitude and enable us to act in an uncaused (or supracausal) way, contributing to our ability to rise above, and overcome, ourselves—even if for a while—revealing our internal connection with that which is not given within the limits of existence. The disinterested stance *straightens* and *rectifies* us, bringing us out of a situation where we are concerned above all with the cares of existence.

In the matter of collecting ourselves, we cannot but be *generous* and even—*extravagant*. It may be a good idea to try and formulate this as a paradox. It may seem to us that we are collecting ourselves when we are involved in the cares of existence, but actually we only waste our energy, i.e. trying to collect ourselves through acquisition, we only expend ourselves and, vice versa, apparently giving away what is our own and disregarding our own interests in unselfishness is the only way we collect ourselves.

Disinterested Acts: The Power of Generosity and of the Gift

Generosity is connected with appreciating the fact that our life is a *gift*. It is given to us, presented to us as a gift, with no compensation expected for the efforts, time and resources that we receive. And this gift is a *primordial* one because everything begins with it. The beginning of a human being is 'from' a gift and 'in' a gift, and as a result, disinterestedness forms the very foundation of our life which has at its centre the action of *giving away* and is the basis for establishing our bond with the disinterested.

[2]It is noteworthy that identity as such is comparable to what is invariable: the Latin *individuum* retains in itself the Greek ἄτομος.

Behind the *miscellaneous*—different objects, different phenomena and different events—there is *one* force at work which does not cease to work even after this miscellany has been formed: everything that has finally taken shape is placed at the epicentre of the action of this force and is constantly in a dynamic condition. The appearances of this force can take different forms: thus, in an anthropological perspective, this force appears as 'passion', whereas in its social dimension, it is seen as a 'process', with people essentially identifying this force with actions and perceiving it as being indissolubly linked with them. One might suggest characterizing such situations of identification as 'traps' and 'conductors' of this force when it is *in ourselves* and *by means of ourselves* that we try to accumulate this force.

At the same time, we understand that without an inflow of energy our strength will be exhausted: thus we encounter, on the one hand, the primacy of this force in relation to us, but on the other hand, we understand that we ourselves are necessary for this force, and without our acceptance of the current of this force our strengthening cannot take place. It is clear that, *without* a conjunction of ourselves and the force operating through us, its impact on us can assume a destructive character. In this case, we find ourselves in a position of someone taken unawares, with the force leading us against our will, as we try to resist it. That is why, being forced to reckon with the impact and influence of this force on us, we have to accept it all at once and unconditionally.

The Gift: Taking and Giving Away. The Act of Self-abandonment

In this regard, understanding the impact of this *force* as a *gift* is capable of changing radically our situation, for, if we let it unfold itself—by means of ourselves and all our strength—then the limits and borders of our possibilities become significantly *extended*. We are now able to expend ourselves practically constantly because our expenditure actually takes the form of a 'gift' and of 'giving away' something that we do not possess, but that we use. That we use as long as we can. And if we do not exceed our limits—the limits of our feelings, the limits of our thinking and those of our will—and if, consequently, we do not find 'channels' for the current of this force, then it will break any limits as if they were barriers standing in its way.

It is necessary to realize that we—ourselves—can become such a barrier. And this is something that often happens: a force in whose way obstacles are placed finds a way out which is unauthorized by us and which now takes an indecent form. Everyone in the course of their life has to face what might be described as an 'overflow' or a 'shortage' of oneself, the consequences of which are psychoses and neuroses. The *extreme* character of such forms is connected precisely with our unwillingness to interact with a vital force which manifests itself in such an extreme way since it aims at reaching, and reaching beyond, the edge (*extremum*) of ourselves. Being a single whole with such a force, we ourselves become its edge.

Thus, we have to do everything we can to let this force unfold itself. For us, this task turns out to be a matter of *giving* all our strength to *another*, whereby an *extension* of the space of our soul takes place. The desire of accumulating and retaining what is one's own within oneself leads to a tragedy, and since we do not, in the ontological sense, possess 'our own' as a form of property, '*our own*' has to be understood not in the parameters of what we have, but precisely in terms of *how* we establish ourselves. In terms of how we establish ourselves through actualization. It is clear that in this respect 'our own' is connected with 'being', not with 'having'.

The habit of perceiving our life as an interaction of *different* and *various* forces which not only fail to be coordinated with each other but also resist each other, must be significantly revised. All of us—both together, and separately—have to experience the action of one force which—in our ignorance—we perceive as *different* types of force. Whether or not to accept the view that behind the perceived variety *one* and the same manifests itself, and that all things separate are a manifestation of this same principle—that is what each of us has to decide one way or the other.

It is necessary to organize oneself in a way that would enable our strength to be channelled into the act of *giving* it to another. Possibly, this is exactly the way we should understand the purpose of our having been called into existence in the world: we come into it because we are *given* strength—for nothing, at once and in its entirety, and this strength is something that we must *give away* and *pass on* to another during our existence. Such an understanding can become central to our life and can give it a sense of direction. Life develops through activity, it will dry up without it.

The Dynamics of the World and the Schematic Attitude. Statements and Records of Facts and the Problem of Understanding

There is another important point connected with the fact that only someone who understands *what* he takes and *for what* he is intended in connection with such a 'burden' can give away. Strength is given to us not only gratis, but also as a loan. In fact, selfishness is a result of one person failing to see the connection of his *own* life with the life of *another*: this is true of both the one who has given or is giving him strength, and the one to whom it has been given. In this sense, man is a bridge. It means being 'between', always and everywhere, between oneself and the other.

If we look at the world from the perspective of strength, then we can see it in constant *dynamics*, whereas any objects that we normally perceive as invariable now appear as changeable pulsations. The permanence of an object is a derivative of its *schematic* consideration, where a scheme makes it possible to pay attention only to what conforms to it and is coordinated with it and, therefore, the dynamic is perceived in an *exclusively* static form. From the angle of a scheme, which embodies an *intellective* attitude to it, the world is considered in an entirely static condition, presupposing in principle the penetration of the intellect into the world, with all that this implies in terms of the growing importance of an *instrumental* and *technological* attitude. The internal movement of the world is eclipsed by the introduction of a scheme: left to our own devices, we can only rely on our own strength, without any inflow of energy from the outside.

At this point, it would appear appropriate to proceed to a brief discussion of what eludes the instrumental and technical attitude. And in this regard, it is worth addressing those layers of language which, apparently, serve to describe a certain state of affairs, while at the same time retaining in themselves something different. A good case in point is the Russian expression for 'as it really is' (*kak ono yest' na samom dele*) which, of course, does more than just refer to a concrete situation. In fact, the latter is relatively unimportant. This expression records the need for the *understanding* of a situation, which implies going beyond the limits of merely recording it. The present situation is directly connected with the need for shifting beyond the limits of *stating* things. In other words, the present embraces *understanding*, whereby it is made more profound and is extended, since something *different*, something *other* than itself has now entered it: a different time, a different place, a different situation.

One could try to describe all this by means of identifying new ways of expressing the similar in the grammatical system of language when the present not only equals the stated but also has a kind of 'residue'. The present 'does not come to an end' at the moment it is recorded, it rather 'lasts' and is 'protracted': it **is** at the moment of its *change*. And it is clear that, seen through such a prism, the possibility of stating things 'as they really are' is no longer a matter of urgency.

The Difficulty of the Present as the Difficulty of Change

Let us pay special attention to the practice of connecting in our consciousness, and even directly identifying the present with recording a certain state of affairs, when the present begins to be perceived as something transient and then increasingly—as a transition from what *was* to what *will be*. As a result, the value of what *is*, i.e. of the present proper, weakens and decreases.

The attitude which accepts the present as such is characteristic of a *creative* dimension which obviously does not coincide with schematism based mainly on intellective understanding. The poet saying 'fiction is not deception'[3] draws our attention to the fact that fiction reveals a different side to what may seem a familiar state of affairs, so that the present comprises itself and something *other* than itself, because otherwise the present can only be considered unilaterally.

In this regard, it is noteworthy that when we persist in upholding the *truth*, we imply that it inevitably involves some kind of interpretation—no matter whether it is ours, his, hers or theirs… The truth is something we are informed about, informed

[3]A quotation from a poem by Bulat Okudzhava (1924–1997).—*Translator's Note.*

by a definite person, which presupposes the inclusion of his understanding in the present: he informs us about something *as* he understands it, and he cannot but transfer to us—in this message—something from *himself.* And this is not a descriptive message, since people would normally speak about truth in a situation which has a strong emotional background whose tension not only fails to be eased but, on the contrary, is maintained as strong as ever.

The fact of the matter is that in our life there always exist certain segments—continents and islands reproducing a creative attitude to what is there when the 'is' can essentially be considered both in a 'cross-section' of the conditions of its maintenance, and in a projection of providing these conditions. It is connected with certain relationships of ours with the 'is' being provided for, and preserved on the part of the other when a mobile area of interrelations between the 'is' and *the other* is established.[4]

Pavel Florensky's Concept of Three Types of Activity. Creativity

It would be interesting in this context to turn to P. A. Florensky's reflections on three types of activity. On his view, man performs actions directed at developing *tools* and *machines*. Producing things in this way actually serves as the basis of *economy.* Along with the creation of tools and mechanisms, there are actions that generate words without which our thinking cannot reveal and realize itself. In the course of such activities *concepts* and *terms* are developed, and a *worldview* is formed. However, considered separately, actions oriented towards producing things, and actions oriented towards thinking are opposed to each other. Apparent is the need for an *embodiment* of sense in life and for a transformation of its material composition thereby. Such an action, according to Florensky, is artistic creativity the pinnacle of which he sees in *devotional activity* in the course of which *the sacred* is 'produced'.[5]

---

[4]It is easy to see that one of the motivations behind M. Heidegger's early philosophy was the reception of such an attitude by modern European philosophy. However it became clear soon enough that it was something fairly difficult to do, and Heidegger decided to make a radical break with the modern European strategy of understanding truth: it is quite possible to interpret in this way all his efforts connected both with the analytic of *Da-sein*, and with grounding ontology in the ontic, as well as with detecting the mobility of the relationships between entity (*Seiendes*) and Being (*Sein*).

[5]See: Svyashchennik Pavel Florensky [Father Pavel Florensky], "Iz bogoslovskogo nalslediya" [From His Theological Heritage], in *Bogoslovskiye trudy* [Theological Works]. Moscow: Izdatel'stvo Moskovsoy Patriarkhii, 1977, pp. 101–109. It is important to mention here that both the production of things—in the form of tools and machines, and the production of words—in the form of concepts and terms, as well as the production of meanings transforming the givenness of the material are connected by Florensky with his view that man produces various organs by means of which, on the one hand, he establishes a connection with life, and on the other hand, his being itself is established.

In addition, coming back to the principal line of these reflections, it should be emphasized that a state of affairs changes already at the time it is recorded. And this change is connected both with an internal change of what has been stated, in general, and in its separate elements, and with a change in the very instance of the stated. The very conditions of stating things change, and they change more and more rapidly, depending on the difficulty of forming our reflexive attitude to it. Our time in general is characterized by a permanent change of the conditions of our attitude—to anything at all—without any effectively formed reflection. As a result, a 'state of affairs' appears not as something that is, but as something thanks to which it arises, without being reflected upon.

The Game as a Means of Resorting to the Immeasurable

In connection with this attitude—which is historically new—to the 'state of affairs', an image inevitably appears which exceeds measure and establishes a possibility of resorting to the *immeasurable*. In other words, some 'device' comes into being enabling *measure* to change and to be filled with the *other*, i.e. to include *another attitude*. It can be clearly seen if one resorts to the image of a game which may assume a variety of forms: a language game or a 'game' of consciousness, even a freak of fate or of chance. In a game-like action definiteness is not only lost, it can also be discovered. In a game we encounter a *development* of definiteness when we can see something that is alive behind the definite. In a game we encounter the *plenitude* of the definite.

It is important that an environment should be reproduced in which it is possible to encounter the plenitude of a 'something' that comprises both itself and a possibility of understanding it, i.e. makes it possible for us to extract a certain sense. If one is able to do so, then the other is revealed as something that is no longer opposed to the definite, but as something immanent in it. The other turns out to be included in, and inherent to understanding.

Reasons and Making Sense of Things

The first step in this direction would probably be to give up *schematism*. We have to accept a basic refusal to recognize the completeness of any schematic description, and therefore to accept the conventional character of the scheme as such and its limitations. Any limits of description are always connected with us: they proceed from our making a pause in understanding things, they proceed from our intention to consider something as exclusively an *object* of our thinking marked by our thought in a way that suggests that this is the only reason for the existence of this 'something'. It is obviously impossible to make any *sense* of things with such an approach: sense is superfluous here and will always be replaced with finding a reason, to be exact—a *reason* will be substituted for it.

In view of the fact that we explain what is going on by means of the concepts 'reason' and 'purpose', our explanation itself is completely identified with our own image. But at the same time, such an explanation is easily applied to any potential object, and even if in giving something the form of an object people sometimes encounter problems, they still try to adhere consistently to just this kind of explanation—one based on objectness.

It is clear that with this approach the *other* as such cannot be perceived: a definite form is suggested for everything, including the other—the form of *objectness*. On the contrary, wherever our relationship with the other is made possible, reasons and purposes become insignificant.

It is certainly possible to assume that the total and permanent *objectivization*, i.e. the activity of giving everything the form of an object, can promote the development of thought. However, this kind of development is actually limited to the establishment of a sphere of *control* in which its components are identified with what can essentially be given an *analytical form* and subjected to change according to certain rules of our thinking.

Thought as the Acquisition of Scope and Extent

It would be highly problematic to speak of *scope* and *extent* of thought within the parameters of objects, as here all our strength is directed at giving the form of an object to everything and at upholding the obvious priority of such a form. Actually, we are dealing here with the sphere of the understanding (*Verstand*) with reference to which it is quite problematic to use the image of scope, as the role of the understanding is fundamentally that of an interface between us and our sensibility. The understanding paves the way for us on which our thought progresses when its movement along the way of expansion in this or that direction has been prohibited, with the result that thinking is actually identified with the activity of moving along a *marked* trajectory from sign to sign.

As was mentioned earlier, we experience a real scope and extent of thought when we are able to accept a relationship with the *other*. The same is true of the scope of our feeling and the extent of our will. The other enables us to go beyond signs that are entirely and completely exhausted in performing their function, so that we can place ourselves in the dimension of symbols.

Signs and Symbols. The Whole

It is important to emphasize that a *symbol*, unlike a *sign*, does not expend itself and does not disappear with use: with a symbol you can always find yet another side to it, yet another aspect, and it is connected with the nature of the other enabling us to go beyond any *limit*. In an orientation towards a symbol and in coordination with it, thought is endowed with a possibility of *free* movement: the symbol sets you free, liberates you from determinism and the identification of thought with purely intellective constructions, when we begin to see the *other* behind the *given*. Whereas signs keep our attention focused only on the definite, the symbol enables us to come into contact with the *Whole* within its *mobile* borders.

The *act of symbolization* is one of the most significant acts of consciousness, as it helps us to break through the limits of the definite and reach the scope of the world itself. It is in this—creative—act that we actually find sense. As far as our attitude to something as a symbol is concerned, everything depends entirely on *us*, we are essentially capable of seeing and understanding any kind of 'something' in an *extended* way when the definite is perceived in the dimension of its establishment. Here we are dealing with the facticity of the whole to which—thanks to a symbol—we are capable of establishing our own attitude. And in this sense symbolization is connected with an act of making 'gaps' in the established order of consideration. It is precisely the *Whole* that gives us scope: it always contains strength in itself and gives us strength. Through the symbolic aspect of the definite, we partake of the Whole, since we penetrate into *another* understanding of *the same*. Thus, symbolic understanding is essentially an *extended* understanding of the world and an understanding of the world in *movement*.

The identification of cause-and-effect links 'works' within strictly localized spatial and temporal coordinates. Taken in itself, each particular definiteness has little to give us: in itself, any definiteness is opposed to other definitenesses, and if it is at all coordinated with them, then again in a strictly definite way. It is extremely important that the symbol—unlike the sign—presupposes a *human* or, more precisely, *our* personal attitude (one which is purely our *own*), whereas the sign is always *conventional* and *impersonal*.

One's Own as the Other

The sign reminds us of that which we, as well as everyone else, irrespective of our and their personal attitude, inevitably have to take into consideration. The symbol presupposes precisely our *own* work of consciousness, when we enter a *direct* and *personal* relationship with a symbol, with all that this implies. If such a relationship with a symbol is not established, nothing occurs, and the symbol is interpreted as a mere sign. The impact of a symbol on us is real and significant only in the case of our direct contact with it. Without establishing a relationship with a symbol which is entirely our *own*, and without disclosing ourselves to the *other* which is a whole, and so—essentially without allowing a possibility of changing ourselves, we identify the symbol with a sign and, therefore, we fail to appreciate the very essence of what we are dealing with.

Thus, the symbol *changes* us, something the sign cannot do. Moreover, being only a point in our reasoning that directs us to itself, the sign *certifies* our givenness. In coordination with a sign, we are not inclined to change: we are *satisfied* with ourselves the way we are. The symbol 'works' if we both accept its openness, and essentially allow our own openness in relation to any possible change of ourselves, something that eventually implies the need for recognizing the *open* character of life in principle.

We should accept that any form of life, no matter how well founded it might seem, is always capable of changing and appearing to us as something different, something other than its former self, therefore—everything is connected either with our ability to accept the *other*, or with our inability to do so. That is why it is so important to cultivate an attitude towards ourselves from a position of the participation of the *other* in our life.

Rejection of Consciousness and the Increased Need for Disinterested Acts

And another important point. For a long time, there existed a paradigm in the parameters of which man essentially relied on a *rational* plane of Being. People's *rational attitude* to things was assumed to be precisely that which enabled them to overcome any difficulty. This 'civilizational', or 'liberal', tradition has a counterpart in another paradigm which involves a re-consideration of the very possibility of rational attitudes on the part of certain people, groups and communities, a paradigm which assumes that not all people can be 'carriers' of a rational attitude. It is clear that in the framework of such a tradition, *violence* becomes a key characteristic of the way people are connected with, and relate to, each other. It is also noteworthy that violence perpetrated on one group of people by another is to some extent 'compensated for' by various forms of violence the first group perpetrates on some other kind of people. One might say that violence becomes a sort of social 'glue' in their life, a situation which, in turn, provides a basis for different ways and forms of life to be converted into, and exchanged for, one another.

Both paradigms, however, presupposed the need for the *figure of the external observer* which could be identified either with some providential power, or with certain sacred centres, or else with qualified specialists. Initially, there was a presumption according to which man could *become conscious* of any moment of his life. In other words, consciousness (thinking)—in its different forms—was perceived as a force unrestricted by anything other than itself, capable of correlating itself with everything under the sun. And even if certain people, groups and communities were seen as susceptible to occasional bouts of 'insanity', which were, however, mainly associated not with 'us', but with 'them' and with the 'others', there was no question of a complete rejection of consciousness. The need for becoming conscious of things presupposed turning to acts of external observation, and required the positing of a transcendental plane of Being.

During periods characterized by a rejection of consciousness, which can be understood as those when consciousness abandons people (and it is obvious that it can leave not only ‘them’, but also ‘us’), the *need* people feel for disinterested acts becomes particularly acute. Such acts can and must now be understood as a means of ‘fine-tuning’ the human in us: in fact, they are exactly that ‘other’ that will, perhaps, be able to make man face his essential Self.

Index

A
Aesthetics, 71, 78, 79
Archetype, 63, 65, 68, 202

B
Białowieża forest, 32
Bodily experience, 197, 198
Border, 15–19, 21–25, 27, 35, 41, 42, 45–49, 51–53, 58, 82–86, 88–92, 95, 130, 135, 138, 149, 155, 157–160, 185, 190, 194, 200, 201, 207, 211, 212, 214, 231
Borderlands, 15–22, 27, 193
Borderology, 18, 41, 42, 44, 46–53, 81–85, 88, 91, 129, 149, 150, 155, 157–160
Border zones, 48, 84, 85, 93, 120
Bounderism, 3, 6

C
Catastrophe, 170
Civilization, 11, 58, 69, 70, 77, 204
Conflict, 31, 34, 36–39, 68, 96, 110, 129, 212, 213
Consciousness, 51, 57, 59, 62, 65–67, 72, 75, 76, 79, 81, 91, 98, 100, 122, 123, 126, 138, 178, 185–193, 198, 199, 202, 231, 239, 241, 243, 244
Cosmopolitan, 81, 83–85, 87, 88, 91, 92, 158
Cosmopolitan rights, 81, 87, 88, 92
Creative thinking, 186, 192
Culture-nature, 3, 4

D
Deconstruction
Deontology, 221, 222, 225
Development, 13, 15, 17, 21–25, 27, 39, 41, 42, 44, 45, 48–51, 53, 64, 67, 69–72, 76, 83, 85, 86, 91, 94, 95, 98, 104, 105, 110, 111, 116, 118, 121, 123, 124, 136–139, 145, 146, 149–152, 154, 157–161, 165, 169, 173, 174, 178, 180–182, 190, 193, 197, 200, 208, 217, 219, 226, 227, 241, 242
Dichotomies, 3, 10

E
Ecological aesthetics, 69, 78, 79
Ecological catastrophe, 70, 77
Ecological economics, 173, 174, 177–182
Ecology, 13, 38, 71, 73, 81, 84, 122, 124, 125, 127, 128, 177, 178, 182
Enlightenment, 4, 69, 73–78, 211
Embeddedness in nature
Environmental identity, 121–127, 130
Environmental protection, x
Environmental migrants, 121, 129, 130
Environmental problems, 73
Eschatology
Epigenetics, 149–153, 157–161
Ethics, 48, 70, 73, 154, 177, 182, 217, 219–222, 225, 227, 228
European culture, 57–60, 62, 63, 65–69, 202
Existence, 5, 13, 45, 46, 50, 51, 58–60, 63, 77, 78, 84, 92, 100, 109, 117, 125, 135, 136, 140, 145, 163, 164, 166, 185–194, 199, 209, 210, 213, 224, 231–236, 238, 241
Experience, 10, 27, 42, 44–46, 48–50, 52, 59, 66, 72, 73, 76–78, 86, 87, 91, 92, 95, 97, 99–101, 114, 115, 120, 143, 151,

J. S. Methi et al. (eds.), *Borderology: Cross-disciplinary Insights from the Border Zone*, Springer Geography,
https://doi.org/10.1007/978-3-319-99392-8

153–155, 157–160, 166–169, 180, 186, 187, 190–194, 198, 199, 203, 204, 209, 210, 213, 217, 219, 224, 225, 227, 228, 232, 233, 235, 238, 242

G
Globalization, 41, 53, 57, 58, 67, 122
Green belt, 84–87, 91, 149
Green economy, 173, 174, 177, 182

H
Hermeneutics, 71, 226, 227
Home, 12, 48, 92, 141, 163, 165, 167–170
Human, 5, 12, 16, 25, 32, 35, 37, 51, 60, 72–76, 79, 85, 87, 88, 91, 92, 94, 97, 115, 121, 123–128, 130, 135–140, 143–146, 151, 154–159, 161, 163–167, 169, 170, 178–180, 185, 186, 188–194, 197–204, 208–211, 214, 219, 226–228, 231–234, 236, 243, 245
Human activity, 73, 139, 149, 151, 154, 156, 157, 159, 160, 166, 197, 210
Husbandry, 3, 7, 139

I
Ideology, 11, 173–178, 181, 182, 202
Images of nature, 197
Imagination, 78, 98, 103, 176, 186, 190, 197–200, 202, 207, 208, 210–214
Interaction, 124, 128, 139–141, 143–146, 149, 150, 152, 156, 158, 159, 166, 181, 185, 238

K
Kola Peninsula, 3, 6, 139

L
Lithuania, 15–25, 27, 32, 41
Living condition, 127, 139, 149

M
Man, 11, 46, 61, 70, 72–76, 78, 81, 82, 84, 86–88, 90–92, 125, 135–141, 145, 146, 155, 156, 166, 178, 180, 182, 185–194, 197, 198, 203, 208, 210–212, 236, 238, 240, 244, 245
Myth, 90, 185–187, 189–194

N
Nature, 3–5, 9, 10, 12, 13, 19, 23, 35–39, 52, 67, 69–73, 78, 81–88, 90–92, 98, 101, 104, 115, 121, 122, 124–128, 135–146, 149, 155, 156, 159, 163–169, 178, 179, 181, 182, 185–194, 197–200, 202–204, 207–209, 213, 219, 226–228, 231, 233, 242
New materialism, 122, 124–126

P
Practical knowledge, 45, 46, 49, 120, 149, 150, 153–160

R
Rationality, 71, 91, 121, 124, 130, 138, 150, 160, 166, 199, 212
Reindeer, 3, 6, 7, 9, 139–142, 144, 189
Russia, 15–25, 27, 33, 42, 43, 46–49, 52, 53, 82, 84, 88, 89, 145, 155, 158, 164, 169, 200, 201, 220, 222, 233

S
Scientific research programs, 49, 62, 64
Social movement
Spatial interaction, 15, 17–19, 21–24, 27
Structure of consciousness
Subsidiarity, 85
Sustainability of culture, 69, 70
Symbol, 31, 89, 166, 187, 188, 190, 231, 242–244
Symbolization, 185, 187–191, 193, 231, 243

T
Technology, 70, 89, 119, 124, 154, 163, 167–169, 194, 203
Translation, 151, 217–228, 233

U
Utopia, 173–178, 181, 182

V
Veer ecology
Virtue ethics, 182, 217, 221, 225, 226, 228

W
Woods (forest), 163, 166, 167

MIX
Papier aus verantwortungsvollen Quellen
Paper from responsible sources
FSC® C105338

If you have any concerns about our products,
you can contact us on
ProductSafety@springernature.com

In case Publisher is established outside the EU,
the EU authorized representative is:
Springer Nature Customer Service Center GmbH
Europaplatz 3, 69115 Heidelberg, Germany

Printed by Libri Plureos GmbH
in Hamburg, Germany